Preface

Welcome to "Robotics: Interview Questions and Answers" – a comprehensive guide designed to navigate the dynamic world of robotics through a lens of inquiry and exploration. In the pages that follow, you will embark on a journey through the fascinating realm of robotics, uncovering a myriad of topics that span the breadth and depth of this transformative field. This book seeks to provide not only a wealth of knowledge but also a practical resource for individuals aspiring to delve into the world of robotics or those seeking to enhance their understanding of its myriad facets. In the age of automation, artificial intelligence, and the Internet of Things, robotics has emerged as a pivotal force shaping our future. From manufacturing floors to healthcare settings, from deep-space exploration to our own living rooms, robots have become an integral part of our daily lives.

Whether you seek to gain knowledge for interviews, academic pursuits, or simply to satisfy your curiosity about the incredible world of robotics, this book is designed to be your trusted companion. It serves as a roadmap to understanding the fundamentals, the nuances, and the future possibilities that robotics holds.

Table of Contents

Chapter 1: Introduction to Robotics

1. What is robotics?

Answer: Robotics is a field of engineering that involves the design, construction, programming, and operation of machines, called robots, to perform tasks autonomously or semi-autonomously.

2. What are the key components of a robot?

Answer: A robot typically consists of three main components: sensors for perception, actuators for movement, and a control system for decision-making.

3. What is the difference between industrial and service robots?

Answer: Industrial robots are used in manufacturing, while service robots are designed for tasks like cleaning, healthcare, and entertainment, often interacting with humans.

4. What is the importance of kinematics in robotics?

Answer: Kinematics studies the motion of robots without considering forces. It's crucial for planning and controlling robot movements accurately.

5. How do robots use sensors for navigation?

Answer: Robots use sensors like LiDAR, cameras, and ultrasonic sensors to perceive their environment and make decisions for navigation.

6. What is the role of artificial intelligence in robotics?

Answer: AI enables robots to learn from data, adapt to changing environments, and make intelligent decisions, making them more versatile and capable.

7. What are the advantages of using robotic automation in manufacturing?

Answer: Robotic automation increases productivity, reduces labor costs, improves product quality, and enhances workplace safety.

8. What is the Uncanny Valley in robotics?

Answer: The Uncanny Valley is a concept where human-like robots that are not quite lifelike can evoke discomfort or eeriness in humans.

9. How do robots communicate with humans?

Answer: Robots can communicate through speech, gestures, touchscreens, and text-based interfaces, enhancing their ability to interact with people.

10. What are the challenges of programming robots for complex tasks?

Answer: Programming robots for complex tasks involves solving issues like path planning, obstacle avoidance, and optimizing task execution.

11. What is the difference between teleoperation and autonomous robots?

Answer: Teleoperation involves humans remotely controlling robots, while autonomous robots operate independently using onboard sensors and algorithms.

12. How do robotic arms mimic human movements?

Answer: Robotic arms use joints and actuators to replicate the seven degrees of freedom found in the human arm, allowing for precise manipulation.

13. What is swarm robotics?

Answer: Swarm robotics involves multiple robots working together collaboratively, inspired by the collective behavior of social insects like ants and bees.

14. How are drones considered a subset of robotics?

Answer: Drones are aerial robots that can be remotely operated or autonomous, often used for tasks like surveillance, mapping, and delivery.

15. What is the role of machine learning in robot perception?

Answer: Machine learning algorithms enable robots to recognize objects, interpret sensor data, and make decisions based on patterns and experience.

16. What are the ethical considerations in robotics?

Answer: Ethical concerns in robotics include issues like privacy, job displacement, and the potential for autonomous weapons.

17. How do robots assist in medical procedures?

Answer: Surgical robots like the da Vinci system enable precision surgeries with minimal invasiveness, reducing patient trauma and recovery time.

18. What is the significance of haptic feedback in robotics?

Answer: Haptic feedback provides robots with a sense of touch, allowing them to interact with objects and environments more effectively.

19. How can robots be used in disaster response?

Answer: Robots can navigate hazardous environments, search for survivors, and deliver supplies during disasters, enhancing safety and efficiency.

20. What are the challenges of developing humanoid robots?

Answer: Creating humanoid robots with natural movements and human-like interaction is challenging due to complex mechanics and control systems.

21. How do robots learn from their mistakes?

Answer: Robots can improve their performance through iterative learning, analyzing failures, and refining their algorithms.

22. What is the role of computer vision in robotics?

Answer: Computer vision enables robots to process and understand visual information, aiding in tasks like object recognition and navigation.

23. How do self-driving cars use robotics technology?

Answer: Self-driving cars employ robotics components like sensors, cameras, and AI algorithms to navigate roads autonomously.

24. What is the concept of "singularity" in robotics?

Answer: The singularity refers to a hypothetical point when robots or AI systems may surpass human intelligence, raising philosophical and ethical questions.

25. How do underwater robots explore the depths of the ocean?

Answer: Submersible robots, equipped with sonar, cameras, and thrusters, investigate ocean environments, collect data, and perform underwater tasks.

26. What are exoskeletons, and how are they used in robotics?

Answer: Exoskeletons are wearable robotic devices that enhance human strength and mobility, aiding individuals with mobility impairments and industrial workers.

27. How do robots enhance the efficiency of agriculture?

Answer: Agricultural robots automate tasks like planting, harvesting, and monitoring crops, increasing yields and reducing labor costs.

28. What is the role of reinforcement learning in robotic control?

Answer: Reinforcement learning allows robots to improve their control strategies by learning from trial and error, adapting to changing conditions.

29. How do robots contribute to space exploration?

Answer: Robots like rovers and drones explore distant planets, collect samples, and conduct experiments, advancing our understanding of the cosmos.

30. What ethical dilemmas arise with the use of autonomous military robots?

Answer: Autonomous military robots raise concerns about the ethics of lethal decision-making without human intervention and the potential for misuse in warfare.

31. How do collaborative robots (cobots) differ from traditional industrial robots?

Answer: Collaborative robots are designed to work alongside humans safely, with sensors and software enabling them to detect and respond to human presence. They facilitate human-robot collaboration in tasks like assembly, inspection, and material handling.

32. What is the concept of swarm intelligence in swarm robotics?

Answer: Swarm intelligence in swarm robotics is inspired by the collective behavior of social insects. It involves a group of robots coordinating their actions through local interactions, enabling them to perform tasks collectively, such as exploration and search missions.

33. How do robots assist in environmental monitoring and conservation efforts?

Answer: Robots equipped with sensors and cameras can monitor ecosystems, track wildlife, and collect data on environmental conditions, aiding in wildlife conservation, and ecological research.

34. What is the role of soft robotics in healthcare?

Answer: Soft robotics involves flexible, compliant materials that mimic natural organisms. In healthcare, soft robots are used for delicate surgeries, rehabilitation, and assistive devices, offering safety and adaptability.

35. How are underwater drones employed in marine research and exploration?

Answer: Underwater drones, or remotely operated vehicles (ROVs), are used to study ocean ecosystems, map the seafloor, and investigate shipwrecks, providing valuable insights into the underwater world.

36. What challenges arise in the development of autonomous flying drones?

Answer: Autonomous drones must overcome obstacles like collision avoidance, battery life, and regulatory compliance to operate safely and efficiently in various applications, such as aerial photography and surveillance.

37. How do robotic exoskeletons benefit rehabilitation therapy?

Answer: Robotic exoskeletons aid individuals with mobility impairments by providing assistance with walking and movement, promoting physical therapy and improving overall quality of life.

38. What is the role of swarm robotics in disaster response?

Answer: Swarm robotics can be used in disaster scenarios for tasks such as search and rescue, debris removal, and hazard assessment, improving the efficiency and safety of emergency response efforts.

39. How are drones used in agriculture for precision farming?

Answer: Agricultural drones equipped with multispectral cameras and sensors can monitor crop health, identify pest infestations, and optimize irrigation, leading to increased crop yields and resource efficiency.

40. How do autonomous delivery robots contribute to last-mile logistics?

Answer: Autonomous delivery robots navigate sidewalks and streets to transport goods, reducing delivery times and costs for businesses, while also decreasing vehicle emissions in urban areas.

41. What is the role of 3D printing in robotics prototyping?

Answer: 3D printing enables rapid and cost-effective prototyping of robot parts and components, allowing engineers to iterate designs quickly and customize robots for specific tasks.

42. How do bio-inspired robots imitate nature's designs?

Answer: Bio-inspired robots mimic the form and function of organisms, such as the gecko's adhesive abilities or a bird's flight, to create robots with enhanced capabilities.

43. What are the ethical considerations in deploying autonomous drones for surveillance?

Answer: The use of autonomous drones in surveillance raises concerns about privacy, data security, and the potential for abuse, necessitating responsible regulations and guidelines.

44. How do modular robots adapt to changing tasks?

Answer: Modular robots consist of interchangeable modules that can be rearranged to perform different tasks, offering flexibility and scalability in robotic systems.

45. What challenges do underwater robots face in deep-sea exploration?

Answer: Deep-sea robots must endure extreme pressure, extreme temperatures, and limited communication, making reliable operation and data collection a significant challenge.

46. How are robots used in the field of space mining?

Answer: Robots are employed for prospecting, excavation, and resource retrieval on celestial bodies like asteroids and the Moon, potentially revolutionizing space resource utilization.

47. What is the role of quantum computing in robotics?

Answer: Quantum computing can accelerate complex robotic simulations and optimization problems, improving robot design and control algorithms.

48. How do drones assist in disaster mapping and response?

Answer: Drones equipped with cameras and LiDAR technology can rapidly survey disaster-stricken areas, providing critical information for rescue and recovery efforts.

49. How can robots be used in the construction industry?

Answer: Construction robots can automate tasks like bricklaying, concrete pouring, and demolition, enhancing efficiency and reducing labor-intensive work.

50. How does biohybrid robotics combine biological and artificial elements?

Answer: Biohybrid robots integrate living tissues or organisms with artificial components, creating robots with biological functions, such as muscle-powered movement.

51. What is the concept of explainable AI (XAI) in robotics?

Answer: Explainable AI in robotics aims to make AI-driven decisions transparent and understandable, enhancing trust and safety in autonomous systems.

52. How do robots contribute to underwater archaeology?

Answer: Underwater robots explore submerged archaeological sites, documenting and preserving historical artifacts and structures hidden beneath the sea.

53. What is the significance of energy-efficient robotics in environmental sustainability?

Answer: Energy-efficient robots reduce power consumption, making them environmentally friendly and suitable for applications like solar-powered agriculture and wildlife monitoring.

54. How do drones aid in wildlife conservation and anti-poaching efforts?

Answer: Drones are used to monitor wildlife populations, detect poaching activities, and protect endangered species by providing real-time surveillance in remote areas.

55. How are robots used in the entertainment industry?

Answer: Robots are employed in entertainment for tasks like animatronics, character performances, and interactive exhibits, creating immersive and engaging experiences.

56. What are the ethical considerations in autonomous military robotics deployment?

Answer: Autonomous military robots raise concerns about the morality of decisions made without human intervention and the potential for lethal actions in warfare, necessitating ethical guidelines.

57. How do robots assist in the inspection and maintenance of critical infrastructure?

Answer: Robots can access and assess infrastructure such as bridges, pipelines, and power lines, reducing the risk to human inspectors and ensuring early detection of issues.

58. What is the role of robotics in the field of agriculture automation?

Answer: Agriculture robots perform tasks like planting, weeding, and harvesting, increasing crop yields and reducing the need for manual labor in farming.

59. How do robots enhance the exploration of extreme environments, like space and deep-sea?

Answer: Robots are designed to withstand extreme conditions and gather valuable data in environments where human presence is challenging or dangerous.

60. What is the future of robotics in healthcare?

Answer: The future of healthcare robotics includes surgical robots, telemedicine, and assistive devices that improve patient care, accessibility, and overall healthcare delivery.

61. What are the applications of swarm robotics in agriculture?

Answer: Swarm robotics aids in precision agriculture by coordinating groups of small robots to perform tasks like planting seeds, applying pesticides, and monitoring soil conditions, leading to more efficient and sustainable farming practices.

62. How do drones assist in disaster relief and humanitarian missions?

Answer: Drones provide crucial aerial assessments, deliver supplies to inaccessible areas, and support search and rescue operations during natural disasters and humanitarian crises, improving response times and outcomes.

63. What role do robots play in the mining industry?

Answer: Robots are used in mining for tasks like drilling, ore extraction, and remote inspection, enhancing safety and efficiency while reducing exposure to hazardous environments.

64. How does bioinformatics intersect with robotics in the field of bio-robotics?

Answer: Bioinformatics aids in designing robots that mimic biological structures and functions, contributing to advancements in fields like prosthetics and medical robotics.

65. What are the challenges in achieving natural language understanding in robots?

Answer: Achieving natural language understanding in robots involves complex language processing, context recognition, and semantic comprehension, requiring advancements in AI and NLP technologies.

66. How can robots enhance education and assist students with special needs?

Answer: Educational robots support personalized learning, provide interactive lessons, and assist students with disabilities, promoting inclusive and effective education.

67. What is the concept of swarming in drone technology?

Answer: Swarming in drone technology involves the coordinated flight of multiple drones, enabling applications like aerial photography, surveillance, and environmental monitoring.

68. How do robots contribute to the exploration of other planets, such as Mars?

Answer: Robots like Mars rovers conduct experiments, collect samples, and explore the Martian surface, advancing our understanding of extraterrestrial environments and the potential for future human missions.

69. How do robots assist in the oil and gas industry for inspection and maintenance?

Answer: Robots equipped with sensors and cameras inspect pipelines, tanks, and offshore platforms, reducing the risk to human workers and ensuring the integrity of critical infrastructure.

70. What is the role of robotic exposits in assisting soldiers and first responders?

Answer: Robotic exposits enhance the physical capabilities of soldiers and first responders, enabling them to carry heavy loads, reduce fatigue, and operate effectively in demanding environments.

71. How can robots be used in the hospitality industry for customer service?

Answer: Robots in hospitality can provide services like room delivery, concierge assistance, and cleaning tasks, improving guest experiences and operational efficiency.

72. What are the potential applications of quantum robotics?

Answer: Quantum robotics may revolutionize tasks like optimization, cryptography, and simulations, enhancing the capabilities of robots in various domains.

73. How do robots assist in underwater archaeology and shipwreck exploration?

Answer: Robots explore underwater archaeological sites and shipwrecks, preserving historical artifacts and uncovering maritime history without disturbing the seabed.

74. What role do robots play in environmental remediation?

Answer: Robots can clean up pollution, remove invasive species, and restore ecosystems, contributing to environmental conservation and restoration efforts.

75. How are drones used in the film and entertainment industry for aerial cinematography?

Answer: Drones equipped with high-quality cameras capture breathtaking aerial shots and dynamic footage for movies, TV shows, and live events, revolutionizing cinematography.

76. What ethical concerns surround the use of autonomous robots in elderly care?

Answer: The use of autonomous robots in elderly care raises questions about privacy, autonomy, and the potential for isolation, necessitating careful consideration of ethical guidelines.

77. How do robots assist in the inspection and maintenance of nuclear facilities?

Answer: Robots with radiation-resistant materials and cameras inspect nuclear facilities, ensuring safety and reducing the exposure of workers to radiation.

78. What is the significance of explainable AI (XAI) in robotic decision-making?

Answer: XAI is crucial for understanding and trusting robotic decisions, particularly in critical applications like autonomous vehicles, where human lives are at stake.

79. How do robots contribute to the field of space tourism?

Answer: Robots assist in spacecraft maintenance, safety protocols, and passenger services, ensuring safe and enjoyable space tourism experiences for future travelers.

80. How can robotics be used in art and creative expression?

Answer: Robotic art installations and creative projects explore the intersection of technology and artistic expression, pushing the boundaries of traditional art forms.

81. What is the role of robotics in the development of smart cities?

Answer: Robotics in smart cities aids in infrastructure maintenance, waste management, and transportation systems, contributing to urban sustainability and efficiency.

82. How do robots assist in the exploration and mining of ocean resources?

Answer: Robots equipped with specialized tools and sensors explore underwater mineral resources, hydrothermal vents, and rare organisms, advancing marine science and resource exploration.

83. What challenges does the integration of AI pose in robotic ethics?

Answer: Integrating AI into robots requires addressing ethical issues like bias, accountability, and decision-making transparency to ensure responsible and ethical use.

84. How do robots support space colonization efforts?

Answer: Robots may be used to build habitats, mine resources, and perform maintenance tasks, laying the groundwork for sustainable human colonies in space.

85. What is the potential impact of quantum sensors in robotic perception?

Answer: Quantum sensors can provide enhanced accuracy and sensitivity to robots, improving their perception abilities for applications like navigation and object detection.

86. How do robots contribute to the field of underwater farming and aquaculture?

Answer: Robots assist in underwater farming by monitoring fish health, maintaining aquaculture infrastructure, and improving the efficiency and sustainability of seafood production.

87. How can robots assist in the detection and removal of landmines?

Answer: Landmine-detection robots equipped with sensors and detectors can identify and safely neutralize landmines, reducing the risks to human definers.

88. What is the role of robotics in the development of autonomous flying taxis?

Answer: Robotics plays a pivotal role in the development of autonomous flying taxis, ensuring safe navigation, passenger comfort, and efficient transportation in urban environments.

89. How are robots used in the fashion industry for design and production?

Answer: Robots contribute to fashion by automating tasks like cutting, sewing, and textile inspection, streamlining production processes and enabling innovative designs.

90. What is the future of swarm robotics in disaster resilience and recovery?

Answer: Swarm robotics will continue to advance disaster resilience by providing efficient and adaptable solutions for search, rescue, and disaster mitigation, saving lives and reducing damage.

91. How do robots assist in deep-sea mining operations for valuable minerals?

Answer: Robots equipped with specialized tools and sensors can extract valuable minerals from the ocean floor, reducing the environmental impact of mining and ensuring resource sustainability.

92. What is the role of robotics in the development of autonomous electric vehicles?

Answer: Robotics technology, such as advanced sensors and self-driving algorithms, is vital in creating safe and efficient autonomous electric vehicles that can reduce traffic congestion and emissions.

93. How do robots contribute to the exploration of active volcanoes and volcanic research?

Answer: Robots equipped with heat-resistant materials and sensors can enter active volcanic environments, collecting data and samples to enhance our understanding of volcanic activity and hazards.

94. What ethical challenges are associated with the use of AI-driven emotional recognition in social robots?

Answer: The ethical concerns involve privacy, consent, and potential misuse of emotional recognition technology in social robots, necessitating ethical guidelines and regulations.

95. How can robots assist in the preservation and restoration of historic buildings and cultural heritage sites?

Answer: Robots can perform delicate tasks like cleaning, restoration, and structural assessments, ensuring the long-term preservation of cultural heritage sites.

96. What is the role of robotics in the development of autonomous agriculture vehicles?

Answer: Autonomous agriculture vehicles equipped with robotics technology can perform tasks like planting, harvesting, and spraying, enhancing crop yields and reducing the need for human labor.

97. How do robotics and AI contribute to the advancement of autonomous underwater exploration?

Answer: Robotics and AI enable autonomous underwater vehicles to navigate and explore deep-sea environments, conduct scientific research, and discover new species and phenomena.

98. What ethical dilemmas arise in the deployment of robotic soldiers and autonomous weaponry?

Answer: Ethical concerns include accountability for actions, civilian safety, and the potential for unintended consequences, leading to debates on the ethics of using robots in warfare.

99. How do robots assist in the rehabilitation of individuals with neurological disorders?

Answer: Robots aid in physical therapy by providing targeted exercises and feedback, helping individuals with neurological disorders regain mobility and function.

100. What is the potential impact of quantum computing on the optimization of robotic algorithms?

Answer: Quantum computing has the potential to revolutionize optimization problems in robotics, allowing for faster and more efficient path planning, resource allocation, and decision-making.

101. How do robots support the mining and extraction of resources on celestial bodies like asteroids?

Answer: Robots are designed for autonomous resource extraction on asteroids, providing a potential source of raw materials for future space exploration and colonization.

102. What is the role of swarm robotics in precision agriculture for crop monitoring and management?

Answer: Swarm robotics enables precise and efficient crop monitoring, pest control, and data collection, improving agricultural productivity while reducing resource usage.

103. How do robots assist in the inspection and maintenance of underwater infrastructure, such as offshore wind farms?

Answer: Robots equipped with underwater cameras and manipulators inspect and maintain offshore infrastructure, ensuring the reliability and efficiency of renewable energy sources.

104. What is the significance of ethical AI in robotic decision-making for autonomous vehicles?

Answer: Ethical AI algorithms are crucial for ensuring that autonomous vehicles make ethical decisions in complex scenarios, prioritizing safety and minimizing harm.

105. How can robots contribute to the recycling and waste management industry?

Answer: Robots can sort, process, and recycle waste materials efficiently, reducing landfill waste and promoting environmental sustainability.

106. What are the challenges and opportunities in the use of swarm robotics for environmental monitoring in forests and jungles?

Answer: Challenges include communication and coordination in dense vegetation, but swarm robotics can provide comprehensive data collection and surveillance in remote ecosystems.

107. How do robots enhance the efficiency of warehouse logistics and order fulfillment?

Answer: Robots in warehouses automate tasks like picking, packing, and transportation, reducing order processing times and optimizing inventory management.

108. What is the role of robotics in assisting individuals with disabilities in their daily lives?

Answer: Assistive robots help individuals with disabilities perform daily tasks like mobility, communication, and self-care, enhancing their independence and quality of life.

109. How do robots assist in the detection and cleanup of marine oil spills?

Answer: Robots equipped with oil-detection sensors and skimming mechanisms can detect and mitigate the impact of marine oil spills on coastal ecosystems.

110. What are the challenges in developing humanoid robots with natural social interaction skills?

Answer: Challenges include developing realistic facial expressions, gestures, and natural language processing for humanoid robots to engage in meaningful social interactions.

111. How can robots be employed in the monitoring and preservation of coral reefs and marine biodiversity?

Answer: Robots equipped with cameras and sensors can survey coral reefs, monitor environmental conditions, and study marine life to support coral reef conservation efforts.

112. What is the potential of robotics in the exploration and colonization of the Moon and Mars?
Answer: Robots are essential for conducting experiments, constructing habitats, and preparing for human missions on the Moon and Mars, advancing our understanding of extraterrestrial environments.
113. How do robots support the field of precision medicine for personalized healthcare?
Answer: Robots assist in precision medicine by automating tasks like drug screening, genetic testing, and targeted therapies, offering customized treatments for patients.
114. What ethical concerns arise in the use of social robots in eldercare and companionship?
Answer: Ethical concerns involve privacy, consent, and emotional manipulation in social robots for eldercare, highlighting the importance of ethical guidelines and oversight.
115. How do robots contribute to the development of sustainable urban agriculture?
Answer: Robots automate tasks like planting, harvesting, and monitoring in urban farms, promoting local food production and reducing the carbon footprint of agriculture.
116. What is the role of robotics in the construction of space habitats for future lunar and Martian colonies?
Answer: Robots assist in constructing and maintaining space habitats, providing the infrastructure necessary for human survival and research on other celestial bodies.
117. How do robots aid in the inspection and maintenance of critical infrastructure like bridges and dams?
Answer: Robots equipped with advanced sensors and climbing mechanisms inspect and repair infrastructure, ensuring public safety and infrastructure longevity.
118. What is the potential impact of AI and robotics on the field of personalized education?
Answer: AI-driven robots can provide personalized learning experiences, adapt curriculum to individual needs, and offer targeted feedback to enhance education outcomes.
119. How do robots contribute to precision aquaculture for sustainable seafood production?
Answer: Robots monitor water quality, feed distribution, and fish health in aquaculture farms, optimizing fish growth and reducing environmental impacts.
120. What ethical considerations surround the use of robotic companions for children and the elderly?
Answer: Ethical concerns include dependency, emotional attachment, and the potential for neglect in robotic companions, necessitating responsible design and usage guidelines.
121. What role do swarm robotics and AI play in autonomous transportation systems like self-driving cars?
Answer: Swarm robotics and AI are integral to autonomous transportation. AI algorithms process data from sensors, while swarm principles aid in vehicle coordination and traffic optimization. These technologies enable self-driving cars to navigate, make decisions, and adapt to dynamic road conditions, enhancing safety and efficiency.
122. How do robots assist in the monitoring and maintenance of offshore renewable energy installations, such as wind turbines and tidal generators?
Answer: Robots designed for offshore maintenance inspect and repair renewable energy infrastructure. Equipped with sensors and manipulators, they can access hard-to-reach areas, reducing maintenance costs and improving the reliability of clean energy sources.
123. What are the challenges and opportunities in using robotics for disaster mitigation and climate change adaptation?
Answer: Challenges include operating in extreme environments and limited communication. However, robots can provide early warning systems, assess damage, and support emergency responses, helping mitigate disaster impacts and monitor climate change effects.
124. How do robots contribute to the development of sustainable farming practices like vertical farming and hydroponics?
Answer: Robots automate tasks in vertical farming and hydroponic systems, such as planting, monitoring, and harvesting crops. This results in increased yields, resource efficiency, and reduced environmental impact.
125. What ethical considerations surround the use of robots in law enforcement and public safety?
Answer: Ethical concerns involve data privacy, surveillance, and potential misuse of force. Striking a balance between security and individual rights is essential when deploying robots in law enforcement.

126. How can robots be used in the exploration and utilization of resources on celestial bodies like the Moon and asteroids?
Answer: Robots equipped with drilling and extraction tools can access and process resources on celestial bodies, supporting future space missions and resource utilization beyond Earth.
127. What role do robots play in the field of remote sensing and environmental monitoring?
Answer: Robots equipped with remote sensing instruments collect data on air and water quality, weather patterns, and ecological changes, aiding in environmental monitoring and research.
128. How do robotics and AI contribute to the development of autonomous air traffic management systems for drones and urban air mobility?
Answer: Robotics and AI facilitate the safe and efficient management of drone traffic in urban environments. They enable autonomous route planning, collision avoidance, and real-time communication among aerial vehicles.
129. What is the potential impact of robotics in the field of neurorehabilitation for individuals with brain injuries or disorders?
Answer: Robots offer repetitive and precise movements in neurorehabilitation therapy. This aids in motor recovery and cognitive rehabilitation for individuals with brain injuries or neurological conditions.
130. How do robots assist in the preservation of natural habitats and wildlife conservation efforts?
Answer: Robots with low environmental impact can monitor and protect natural habitats, track wildlife populations, and combat illegal poaching, contributing to wildlife conservation.
131. What ethical concerns arise in the use of robots for eldercare and healthcare, particularly in end-of-life care?
Answer: Ethical dilemmas involve consent, dignity, and emotional support in end-of-life care. Balancing technology and human touch is crucial to ensure ethical use of robots in healthcare.
132. How can robots support the restoration and maintenance of historical monuments and architectural heritage?
Answer: Robots equipped with precision tools can restore delicate architectural elements, conduct structural assessments, and clean historical monuments, preserving cultural heritage.
133. What is the role of robotics in the exploration and study of extreme environments, such as deep-sea hydrothermal vents and Arctic regions?
Answer: Robots equipped for extreme environments collect data, study unique ecosystems, and conduct experiments in regions where humans cannot easily access, expanding our scientific knowledge.
134. How do robots contribute to the development of clean and renewable energy sources, such as solar and wind power?
Answer: Robots maintain and clean solar panels, inspect wind turbines, and optimize renewable energy installations, enhancing energy generation and reducing maintenance costs.
135. What ethical considerations surround the use of autonomous drones in surveillance, privacy, and civil liberties?
Answer: Ethical concerns pertain to surveillance abuses, privacy infringements, and potential misuse of autonomous drones. Striking a balance between security and civil liberties is essential.
136. How can robots assist in the detection and management of invasive species in ecosystems?
Answer: Robots equipped with sensors and recognition algorithms can detect and remove invasive species, preventing ecological damage and preserving native biodiversity.
137. What is the potential for robots in disaster-resilient and sustainable infrastructure design and construction?
Answer: Robots can construct resilient infrastructure, monitor structural health, and respond to disasters, enhancing the safety and longevity of critical infrastructure.
138.How do robots contribute to the field of telemedicine and remote healthcare, especially in underserved areas?
Answer: Telemedicine robots enable remote medical consultations, monitoring, and diagnostics, bridging healthcare gaps in remote and underserved regions.

139. What are the ethical implications of using robots for military purposes, including autonomous weapon systems?
Answer: Ethical concerns revolve around the use of lethal autonomous weapons without human intervention. International agreements and ethical principles are necessary to address these issues.
140. How can robots be employed in the field of precision agriculture for soil analysis and nutrient management?
Answer: Robots equipped with soil sensors can analyze nutrient levels and provide precise recommendations for fertilizer application, reducing waste and promoting sustainable agriculture.
141. What role do robots play in the exploration and study of extreme climates, such as polar ice caps and desert environments?
Answer: Robots equipped with environmental sensors can gather data in extreme climates, helping researchers study climate change, conduct geological surveys, and monitor ecosystem health.
142. How do robots contribute to the development of smart and energy-efficient buildings?
Answer: Robots can automate building management tasks, such as HVAC control and energy optimization, improving energy efficiency and reducing environmental impact.
143. What is the role of robotics in the field of assisted living and healthcare for the elderly population?
Answer: Robots assist with tasks like medication reminders, mobility support, and companionship for elderly individuals, enhancing their independence and quality of life.
144. How can robots aid in the management of water resources and environmental conservation efforts in aquatic ecosystems?
Answer: Robots equipped with water quality sensors can monitor and manage water resources, support aquatic biodiversity, and mitigate pollution in aquatic ecosystems.
145. What ethical concerns arise in the use of robots in education and personalized learning, particularly in terms of data privacy and student autonomy?
Answer: Ethical dilemmas involve data privacy, surveillance, and the potential for manipulation in personalized learning. Safeguarding student data and respecting autonomy are essential considerations.
146. How do robots contribute to the exploration and study of extreme underground environments, such as caves and subterranean ecosystems?
Answer: Robots equipped with cameras, sensors, and mobility capabilities can navigate underground environments, document cave formations, and study subterranean life forms, expanding our knowledge of Earth's hidden realms.
147. What is the potential impact of robotics and AI in the field of drug discovery and pharmaceutical research?
Answer: Robotics and AI enable high-throughput screening of compounds, accelerating drug discovery and reducing the time and cost of bringing new pharmaceuticals to market.
148. How do robots assist in the inspection and maintenance of critical infrastructure in densely populated urban areas?
Answer: Robots equipped for urban infrastructure inspection can assess bridges, tunnels, and utilities, ensuring public safety and preventing service disruptions in crowded city environments.
149. What role do robots play in the development of autonomous underwater research vessels (AUVs) for ocean exploration and data collection?
Answer: Robots are integral to AUVs, enabling them to navigate, collect oceanographic data, and conduct scientific experiments in remote and deep-sea environments.
150. What is the significance of robotics in the field of space manufacturing and resource utilization for future space colonies?
Answer: Robots will be crucial for building structures, processing resources, and supporting life on space colonies, enabling sustainable human presence beyond Earth.
151. How can robots contribute to the reduction of plastic waste in oceans and water bodies?
Answer: Robots designed for marine cleanup can locate, collect, and recycle plastic waste from oceans and water bodies, mitigating the environmental impact of plastic pollution.

152. What role do robots play in the field of disaster prediction and early warning systems, such as earthquake and tsunami monitoring?
Answer: Robots equipped with seismic sensors and data analysis capabilities aid in earthquake prediction and tsunami monitoring. They provide valuable data for early warning systems, enhancing disaster preparedness.
153. How do robots assist in the restoration of degraded ecosystems and reforestation efforts?
Answer: Robots can plant trees, restore damaged landscapes, and combat deforestation, aiding in the restoration of ecosystems and addressing climate change.
154. What is the potential for robotics in the development of sustainable and autonomous farming on Mars and other extraterrestrial environments?
Answer: Robots are critical for cultivating crops and managing resources in extraterrestrial farming, supporting future human colonization and resource utilization.
155. How can robotics and AI contribute to the development of autonomous marine vessels for scientific research and ocean exploration?
Answer: Robotics and AI enable autonomous marine vessels to conduct scientific research, map ocean depths, and monitor marine life, advancing our understanding of the world's oceans.
156. What ethical concerns arise in the use of robots for autonomous decision-making in legal and judicial processes?
Answer: Ethical dilemmas include accountability, transparency, and bias in autonomous legal processes. Ensuring fairness and adherence to legal principles is essential when employing robots in the legal system.
157. How do robots contribute to the field of precision medicine for cancer diagnosis and treatment?
Answer: Robots assist in precise surgeries, deliver targeted therapies, and analyze large datasets in cancer research, improving patient outcomes and personalized treatments.
158. What is the role of robotics in the development of autonomous flying drones for environmental monitoring and wildlife conservation?
Answer: Robotics technology enables autonomous drones to monitor and protect wildlife, assess habitat health, and combat illegal poaching, supporting biodiversity conservation efforts.
159. What ethical considerations surround the use of robots in journalism and media, particularly in generating news content and deepfake prevention?
Answer: Ethical concerns include misinformation, privacy violations, and deepfake creation. Responsible journalism ethics and technology safeguards are essential in robot-generated news content.
160. How can robots be employed in the field of art therapy and creative expression for individuals with mental health challenges?
Answer: Robots can assist in art therapy by providing creative tools, emotional support, and a non-judgmental environment, aiding individuals in managing mental health challenges.
161. What is the potential for robotics in the development of autonomous underground transportation systems, such as tunnel boring and subterranean transportation networks?
Answer: Robots play a crucial role in tunnel boring and maintenance for underground transportation systems, enabling efficient and safe subterranean travel and infrastructure development.
162. How do robots assist in the exploration and study of extreme environments on Earth, like active volcanoes and deep-sea hydrothermal vents?
Answer: Robots equipped with specialized sensors and materials can collect data and samples from extreme environments, enhancing our understanding of these unique ecosystems.
163. What ethical concerns arise in the use of social robots in eldercare and long-term care facilities, particularly regarding the emotional well-being of residents?
Answer: Ethical concerns include loneliness, attachment, and the potential for neglect in eldercare settings using social robots. Ethical guidelines should prioritize residents' emotional well-being and quality of life.
164. How can robots support the conservation of endangered species through assisted reproduction and breeding programs?

Answer: Robots can assist in assisted reproduction techniques for endangered species, such as artificial insemination and embryo transfer, helping to increase genetic diversity and prevent extinction.

165. What is the role of robotics in the development of autonomous underwater habitats and research stations for deep-sea exploration?

Answer: Robots contribute to the construction, maintenance, and operation of autonomous underwater habitats, supporting scientific research and human exploration of deep-sea environments.

166. How do robots assist in the management and maintenance of large-scale agricultural operations, such as agribusinesses and plantations?

Answer: Robots automate tasks in large-scale agriculture, such as planting, harvesting, and irrigation, optimizing resource use and crop production for agribusinesses.

167. What ethical considerations surround the use of robots for surveillance in public spaces and urban environments?

Answer: Ethical concerns involve privacy, surveillance, and potential misuse of data in urban surveillance. Responsible use of surveillance robots should prioritize privacy and civil liberties.

168. How can robots contribute to the development of sustainable and energy-efficient transportation systems, including electric vehicles and public transit?

Answer: Robots assist in the manufacturing and maintenance of electric vehicles, optimizing energy efficiency and supporting sustainable transportation options.

169. What is the potential for robotics in the field of personalized fashion and clothing manufacturing?

Answer: Robots can enable personalized clothing manufacturing by customizing designs, patterns, and sizes, reducing waste in the fashion industry and promoting sustainability.

170. How do robots aid in the detection and mitigation of air pollution and environmental contaminants in urban areas?

Answer: Robots equipped with air quality sensors can monitor pollution levels, identify sources, and support efforts to reduce air pollution and protect public health in urban areas.

ONLINE RESOURCES

Practice Paper Link: https://forms.gle/a5uFXCdhLDFa7QoZ9

QR Code :

Video Content Link: https://www.youtube.com/playlist?list=PLzlh_31VWEFr8ug6QR9xfGYbQqDxURTbI

QR Code :

Chapter 2: Robot Components and Architecture

1. What is the primary purpose of robot sensors?

Answer: Robot sensors detect and collect data about the robot's environment, enabling it to make informed decisions and interact with the surroundings.

2. How do actuators contribute to a robot's functionality?

Answer: Actuators are responsible for executing movements and actions in a robot, such as turning wheels, moving arms, or manipulating objects.

3. What is the significance of a robot's controller or microcontroller?

Answer: The controller serves as the robot's brain, processing sensor data and sending commands to actuators, allowing it to perform tasks.

4. What are the main types of robot sensors?

Answer: Common robot sensors include proximity sensors, vision sensors, temperature sensors, and touch sensors.

5. How do proximity sensors work in robotics?

Answer: Proximity sensors detect the presence or absence of objects by emitting a signal and measuring the reflection or absorption of that signal.

6. What is the role of vision sensors in robots?

Answer: Vision sensors provide robots with the ability to perceive and understand their surroundings through image processing and object recognition.

7. How do temperature sensors contribute to robot functionality?

Answer: Temperature sensors measure the temperature of the robot's environment or components, ensuring safe operation and enabling thermal management.

8. What are touch sensors used for in robotics?

Answer: Touch sensors detect physical contact or pressure changes, allowing robots to interact with objects and respond to touch.

9. What is the purpose of a robot's manipulator or robotic arm?

Answer: The manipulator or robotic arm is responsible for precise and controlled movements, enabling tasks such as picking up objects or performing surgeries.

10. How do robot grippers and end-effectors vary in their functions?

Answer: Grippers are used to grasp and hold objects, while end-effectors are specialized attachments that perform specific tasks, such as welding or 3D printing.

11. What is the role of a robot's power supply or battery?

Answer: The power supply provides energy for the robot's operation, allowing it to function autonomously or with external power sources.

12. How do mobility systems in robots impact their movement capabilities?

Answer: Mobility systems determine how robots move, whether on wheels, tracks, legs, or aerial platforms, influencing their adaptability to different environments.

13. What is the significance of a robot's communication interface?

Answer: Communication interfaces enable robots to exchange data with external devices, networks, or human operators, facilitating control and data sharing.

14. How do robot processors affect their computational capabilities?

Answer: The processor's speed and capabilities determine a robot's ability to process data and make decisions, affecting its overall performance.

15. What are the key components of a robot's architecture?

Answer: A robot's architecture typically includes sensors, actuators, a controller, power supply, mobility system, communication interface, and processors.

16. How do robots utilize feedback control systems?

Answer: Feedback control systems use sensor data to adjust and optimize a robot's actions, ensuring accuracy and stability in various tasks.

17. What is the significance of redundancy in robot design?
Answer: Redundancy provides backup systems or components, enhancing a robot's reliability and fault tolerance.
18. How do robots utilize algorithms for navigation and decision-making?
Answer: Robots employ algorithms to process sensor data, plan routes, and make decisions, enabling autonomous movement and task execution.
19. What is the primary function of a robot's onboard memory?
Answer: Onboard memory stores data, instructions, and software necessary for a robot's operation and task execution.
20. How do robots adapt to dynamic environments?
Answer: Robots can adapt to changing environments through sensor feedback, real-time data analysis, and decision-making algorithms.
21. What are the advantages of modular robot design?
Answer: Modular design allows for flexibility, easy maintenance, and the ability to upgrade or reconfigure robot components as needed.
22. What are the limitations of fixed architecture robots?
Answer: Fixed architecture robots may lack adaptability and struggle to perform tasks beyond their original design specifications.
23. How do robots with distributed architectures function?
Answer: Robots with distributed architectures delegate control and processing tasks to multiple components, improving scalability and fault tolerance.
24. What is the significance of real-time operating systems (RTOS) in robot control?
Answer: RTOS ensures timely execution of critical tasks, making it essential for applications where precise timing and responsiveness are crucial.
25. How do safety features, such as emergency stop buttons, enhance robot operation?
Answer: Safety features provide a quick means to halt robot operation in emergencies, preventing accidents and damage.
26. What is the role of human-machine interfaces (HMIs) in robot control?
Answer: HMIs enable human operators to interact with and control robots through intuitive interfaces, enhancing ease of use and safety.
27. How do robots utilize feedback loops in control systems?
Answer: Feedback loops compare desired and actual outcomes, allowing robots to adjust their actions for improved accuracy and performance.
28. What are the advantages of autonomous robots?
Answer: Autonomous robots operate without constant human intervention, reducing labor costs and enabling tasks in hazardous or remote environments.
29. How do collaborative robots (cobots) differ from traditional robots?
Answer: Cobots are designed to work alongside humans safely, facilitating human-robot collaboration and teamwork.
30. What are the key considerations in selecting robot components for a specific application?
Answer: Factors such as the environment, task complexity, precision requirements, and safety concerns influence the choice of robot components for a particular application.
31. What are the primary challenges in designing robot sensors for extreme environments, such as space or deep-sea exploration?
Answer: Sensors for extreme environments must withstand extreme temperatures, pressures, and radiation while providing accurate data. Developing materials and technologies that can endure these conditions is a significant challenge.
32. How do robots with multi-modal sensors integrate data from different sensor types to enhance perception?

Answer: Multi-modal sensor fusion combines data from various sensors, like cameras and LiDAR, to create a more comprehensive and accurate representation of the robot's environment, improving perception and decision-making.

33. What role do embedded systems play in controlling robot components like sensors and actuators?

Answer: Embedded systems manage real-time communication between sensors, actuators, and the robot's controller, ensuring coordinated actions and efficient operation.

34. How do soft robotics differ from traditional rigid robotics in terms of design and applications?

Answer: Soft robotics use flexible materials and air or fluid pressure for movement, making them suitable for delicate tasks and applications requiring interaction with humans.

35. What are the benefits of energy-efficient actuators in robotics, especially for autonomous mobile robots?

Answer: Energy-efficient actuators prolong battery life in mobile robots, allowing for longer missions and reducing the need for frequent recharging.

36. How do robots use feedback control to maintain stable and accurate movements during tasks like pick-and-place operations?

Answer: Feedback control systems continuously compare the robot's actual position with the desired position and make adjustments to ensure precise movements, essential for tasks requiring accuracy.

37. What are the advantages of parallel robot architectures, such as delta robots, in high-speed and precision applications?

Answer: Parallel robots offer high-speed and precision due to their reduced inertia and rigid structure, making them ideal for tasks like pick-and-place operations and 3D printing.

38. How do swarm robotics systems enable collaborative behaviors among multiple robots to accomplish complex tasks?

Answer: Swarm robotics systems rely on local communication and coordination rules to enable multiple robots to work together, allowing them to perform tasks collectively that would be challenging for a single robot.

39. What are the key considerations in selecting the appropriate power supply for a robot, especially for remote or off-grid applications?

Answer: Considerations include energy capacity, weight, and recharging options. In remote or off-grid scenarios, power supplies need to provide sufficient energy for the robot's mission duration.

40. How do humanoid robots mimic human movements and behaviors, and what are their applications?

Answer: Humanoid robots imitate human movements using multiple joints and sensors. They have applications in healthcare, entertainment, and research, including human-robot interaction studies.

41. What is the role of redundancy in robot control systems, and how does it enhance reliability?

Answer: Redundancy involves incorporating backup components or systems into robot control. In case of component failure, redundancy ensures continued operation, enhancing overall system reliability. For example, redundant actuators in a robot's arm allow it to perform tasks even if one actuator malfunctions.

42. How do robots with machine learning capabilities adapt to changing environments and tasks?

Answer: Robots equipped with machine learning algorithms can adapt to changing environments and tasks by continuously analyzing data and adjusting their behavior. This enables them to learn and improve over time without explicit programming.

43. What challenges are associated with ensuring the safety of autonomous robots, especially in shared spaces with humans?

Answer: Challenges include collision avoidance, real-time risk assessment, and developing robust safety mechanisms. Ensuring safety is critical to prevent accidents and maintain trust in robot-human interactions.

44. How do haptic feedback systems in robots enhance human-robot interaction and teleoperation?

Answer: Haptic feedback systems provide tactile sensations to users, improving their sense of touch and control when operating robots remotely. This enhances teleoperation precision and situational awareness.

45. What are the implications of using robots in surgical procedures, and how do they contribute to minimally invasive surgery?

Answer: Surgical robots offer precision and dexterity, reducing the invasiveness of procedures. They can access tight spaces and provide magnified views, minimizing patient trauma and improving surgical outcomes.

46. What are the key considerations in designing robot mobility systems for challenging terrains, such as rough terrain or disaster-stricken areas?

Answer: Mobility systems for challenging terrains should prioritize stability, adaptability, and robustness. Features like adjustable suspension and advanced traction control are crucial for navigation.

47. How do underwater robots, like remotely operated vehicles (ROVs), contribute to deep-sea exploration and research?

Answer: ROVs equipped with cameras and sensors can reach extreme depths, allowing scientists to explore and study deep-sea ecosystems, hydrothermal vents, and archaeological sites.

48. What role do drones and aerial robots play in applications like agriculture, search and rescue, and surveillance?

Answer: Drones and aerial robots provide aerial perspectives for tasks such as crop monitoring, disaster response, and surveillance. Their agility and mobility make them valuable assets in various industries.

49. How do robots assist in the inspection and maintenance of critical infrastructure, such as bridges and pipelines?

Answer: Robots equipped with specialized sensors can inspect infrastructure for defects and wear, ensuring safety and reducing the need for costly human inspections.

50. What ethical considerations surround the use of autonomous robots in decision-making, such as in autonomous vehicles and medical diagnosis?

Answer: Ethical dilemmas include liability, accountability, and moral decision-making in autonomous systems. Establishing ethical guidelines and responsible AI is essential to address these concerns.

51. What is the role of robots in the field of precision agriculture, and how do they optimize resource use in farming?

Answer: Robots in precision agriculture use data from sensors and GPS technology to precisely manage planting, irrigation, and harvesting, reducing resource waste and improving crop yields.

52. How do robotic exoskeletons enhance mobility and rehabilitation for individuals with mobility impairments?

Answer: Robotic exoskeletons provide support and assist with leg movements, enabling individuals with mobility impairments to walk and engage in physical therapy for rehabilitation.

53. What challenges and opportunities are associated with the development of autonomous drones for package delivery and logistics?

Answer: Challenges include airspace regulations and safety concerns, but autonomous drones can revolutionize package delivery by increasing speed and efficiency.

54. How can robots contribute to the monitoring and maintenance of renewable energy sources, such as solar farms and wind turbines?

Answer: Robots equipped with sensors and cleaning mechanisms can maintain and optimize renewable energy installations, increasing energy generation efficiency.

55. What is the significance of swarm robotics in applications like environmental monitoring and disaster response?

Answer: Swarm robotics enables multiple robots to collaborate and cover larger areas in environmental monitoring and disaster response, providing more comprehensive data and quicker response times.

56. How do robots assist in the inspection and repair of underwater infrastructure, including offshore oil rigs and underwater pipelines?

Answer: Underwater robots equipped with cameras and manipulators can access and inspect underwater infrastructure, making repairs and ensuring safety in challenging underwater environments.

57. What are the key considerations in designing robots for space exploration, including factors like extreme temperatures and radiation?

Answer: Space robots must be radiation-hardened, thermal-resistant, and equipped with advanced autonomy to navigate and conduct experiments in harsh space environments.

58. How do robots contribute to the development of smart cities and urban planning, including traffic management and waste collection?

Answer: Robots play a role in smart cities by optimizing traffic flow, collecting data for urban planning, and managing waste collection, contributing to more efficient and sustainable urban environments.

59. What ethical challenges arise in the use of autonomous military robots, especially in combat situations?

Answer: Ethical concerns include autonomous weapon systems and the lack of human judgment in combat decisions. International agreements and ethical principles are necessary to address these issues.

60. How do robots with artificial intelligence capabilities assist in language translation and communication in multilingual settings?

Answer: AI-driven robots can provide real-time language translation, breaking down communication barriers in multilingual environments, and enhancing cross-cultural interactions.

61. What is the role of robots in assisting individuals with autism and neurodevelopmental disorders, particularly in therapy and social skills development?

Answer: Robots can provide structured and predictable interactions to help individuals with autism and neurodevelopmental disorders practice social skills and improve communication.

62. How do robots contribute to the field of industrial automation and manufacturing, and what are the advantages of collaborative robots (cobots)?

Answer: Robots in industrial automation perform tasks like assembly and welding, improving efficiency and quality. Cobots work alongside humans, enhancing safety and flexibility in manufacturing.

63. What challenges and opportunities are associated with the use of humanoid robots in healthcare, including roles in rehabilitation and patient care?

Answer: Challenges include cost and complexity, but humanoid robots have the potential to provide personalized rehabilitation exercises and support for healthcare professionals in patient care.

64. How do robots with artificial intelligence algorithms assist in financial trading and investment strategies?

Answer: AI-powered trading robots analyze market data and execute buy or sell orders based on predetermined algorithms, enhancing trading efficiency and decision-making.

65. What is the role of robotics in the field of entertainment and theme park attractions, and how do they enhance visitor experiences?

Answer: Robotics create interactive and immersive experiences in theme parks through animatronics, ride systems, and character interactions, providing entertainment value for visitors.

66. How can robots assist in the exploration and colonization of extraterrestrial bodies, such as Mars and the Moon?

Answer: Robots are essential for conducting experiments, constructing habitats, and preparing for human missions on the Moon and Mars, advancing our understanding of extraterrestrial environments.

67. What is the significance of swarm robotics in agriculture, particularly in tasks like pollination and crop monitoring?

Answer: Swarm robots can mimic natural behaviors, such as pollination by bees, to assist in crop fertilization. Additionally, they can monitor crop health and provide data-driven insights for precision agriculture, optimizing yield and resource usage.

68. How do robots equipped with advanced vision systems contribute to quality control and inspection processes in manufacturing industries?

Answer: Robots with vision systems can detect defects and anomalies in products with high precision and speed. They ensure consistent quality in manufacturing by identifying flaws that may be missed by human inspectors.

69. What ethical concerns arise in the use of social robots for companionship and emotional support, especially for vulnerable populations like the elderly?

Answer: Ethical considerations involve privacy, emotional attachment, and the potential for depersonalization in care settings. Striking a balance between human interaction and robot companionship is crucial for ethical use.

70. How can robots aid in archaeological excavations and the preservation of historical artifacts?

Answer: Robots equipped with sensors and manipulators can excavate archaeological sites delicately and document artifacts precisely, preserving historical treasures while providing valuable insights for researchers.

71. What role do robots play in space mining and resource utilization on celestial bodies?

Answer: Robots are crucial for mining and processing resources on celestial bodies like asteroids, providing raw materials for space colonies and future space exploration missions.

72. How do robots assist in disaster recovery and relief efforts, including search and rescue operations in hazardous conditions?

Answer: Robots can navigate hazardous environments, locate survivors, and deliver supplies during disaster response. Their deployment enhances the efficiency and safety of rescue operations.

73. What ethical considerations surround the use of robots in elderly care, especially in terms of privacy, autonomy, and dignity?

Answer: Ethical concerns involve privacy infringements, decision-making autonomy, and maintaining the dignity of elderly individuals. Ethical guidelines should prioritize the well-being and preferences of elderly care recipients.

74. How do robots with natural language processing capabilities contribute to customer service and virtual assistants?

Answer: Robots with natural language processing can interact with customers, answer questions, and assist with tasks, enhancing customer service and user experiences in various industries.

75. What is the potential impact of robotics in the field of art and creative expression, including robot-generated art and interactive installations?

Answer: Robots have the potential to create unique and innovative forms of art, engaging audiences through interactive installations and collaborations with artists.

76. How can robots aid in the restoration and maintenance of underwater cultural heritage sites, such as shipwrecks and submerged archaeological sites?

Answer: Robots can access underwater cultural heritage sites, document their condition, and perform preservation and restoration tasks without disturbing the delicate ecosystems and artifacts.

77. What is the role of robotics in the field of environmental conservation, particularly in combating deforestation and habitat restoration?

Answer: Robots can plant trees, monitor deforestation, and restore habitats efficiently, aiding in the conservation of biodiversity and natural ecosystems.

78. How do robots equipped with advanced materials and mechanisms contribute to disaster-resilient infrastructure design and construction?

Answer: Robots can construct earthquake-resistant buildings, monitor structural health, and respond to disasters promptly, contributing to the creation of resilient and safe infrastructure.

79. What ethical concerns arise in the use of robots for surveillance and law enforcement, especially in terms of privacy and data security?

Answer: Ethical dilemmas include the potential for mass surveillance, data breaches, and privacy infringements. Safeguarding privacy and data security is paramount when deploying robots in law enforcement.

80. How can robots support the development of sustainable transportation systems, including electric and autonomous vehicles, to reduce environmental impact?

Answer: Robots aid in the manufacturing and maintenance of electric and autonomous vehicles, accelerating their adoption and reducing greenhouse gas emissions in transportation.

81. What role do robots play in the exploration and study of extreme environments on other planets, such as Venus and Europa?

Answer: Robots equipped for extreme environments can explore other planets, collect data, and search for signs of life, advancing our understanding of the solar system's unique locales.

82. How do robots assist in the inspection and maintenance of critical infrastructure in remote and hard-to-reach locations, like oil rigs in the open sea?

Answer: Robots can inspect and repair infrastructure in remote areas, such as offshore oil rigs, reducing the need for human workers in challenging and hazardous conditions.

83. What is the potential for robotics in the development of autonomous agricultural machinery and precision farming?

Answer: Robotics technology can automate agricultural tasks, enhance crop monitoring, and enable precision farming practices, increasing agricultural efficiency and sustainability.

84. How do robots with advanced sensors and artificial intelligence contribute to wildlife conservation efforts, including anti-poaching initiatives?

Answer: Robots equipped with sensors and AI can detect illegal poaching activities, monitor wildlife populations, and assist in conservation efforts, protecting endangered species.

85. What are the ethical implications of using robots for education, including the potential for data privacy breaches and technological inequality?

Answer: Ethical concerns involve data privacy, access to technology, and potential biases in educational robots. Ensuring equitable access and responsible data handling is essential in educational robotics.

86. How can robots be employed in the field of climate science, including data collection and monitoring of environmental changes?

Answer: Robots equipped with environmental sensors and mobility capabilities can collect climate data, monitor changes in ecosystems, and contribute to climate research efforts.

87. What is the significance of robotics in the development of autonomous underwater research vessels (AUVs) for ocean exploration and data collection?

Answer: AUVs equipped with robotics technology can explore and map the ocean depths, collect oceanographic data, and conduct scientific experiments in remote and challenging marine environments.

88. How do robots contribute to the field of neuroscience research and understanding of the brain's functions?

Answer: Robots assist in neuroscientific experiments by providing precise movements and controlled stimuli, aiding researchers in studying brain functions and disorders.

89. What is the potential impact of robotics in the field of agricultural pest control and sustainable pest management practices?

Answer: Robots can target specific pests with precision, reducing the need for pesticides and promoting sustainable pest management practices in agriculture.

90. How can robots assist in the development of smart cities, particularly in urban transportation and traffic management?

Answer: Robots can optimize traffic flow, assist with public transportation, and collect data for smart city planning, contributing to efficient and sustainable urban transportation systems.

91. What is the role of robotics in the field of nuclear decommissioning and radioactive waste management?

Answer: Robots are crucial for safely accessing and dismantling nuclear facilities, handling radioactive materials, and managing radioactive waste, reducing human exposure to radiation.

92. How can robots aid in the detection and removal of landmines and unexploded ordnance, contributing to post-conflict recovery and safety?

Answer: Robots equipped with sensors and mine-clearing mechanisms can detect and safely remove landmines and unexploded ordnance, enabling land rehabilitation and protecting communities.

93. What ethical considerations arise in the use of robots for military applications, including autonomous weapon systems and drone warfare?

Answer: Ethical concerns include the potential for loss of human control, civilian casualties, and the proliferation of autonomous weaponry. Ethical guidelines and international agreements are essential to address these issues.

94. How do robots with natural language understanding capabilities enhance human-robot interaction in customer service and virtual assistant applications?

Answer: Robots with natural language understanding can engage in more nuanced and context-aware conversations with users, improving customer service quality and virtual assistant performance.

95. What is the role of robotics in space colonization and the establishment of sustainable human habitats on other celestial bodies?
Answer: Robots play a pivotal role in building infrastructure, resource extraction, and habitat construction for sustainable human colonization beyond Earth.
96. How do robots with advanced mobility systems assist in the inspection and maintenance of large-scale infrastructure, such as bridges and dams?
Answer: Robots with mobility systems can access and inspect hard-to-reach infrastructure, perform maintenance tasks, and prevent structural failures, ensuring public safety.
97. What are the ethical implications of using robots in the military for surveillance and reconnaissance, especially in terms of privacy and data security?
Answer: Ethical concerns include privacy infringements, data security breaches, and potential misuse of surveillance capabilities. Responsible use of military robots must prioritize privacy and data protection.
98. How can robots contribute to the field of telemedicine and remote medical consultations, particularly in improving healthcare access for underserved populations?
Answer: Robots can facilitate telemedicine by providing remote consultations, enabling healthcare access in rural and underserved areas, reducing healthcare disparities.
99. What is the significance of robotics in the development of advanced prosthetic limbs and assistive devices for individuals with limb loss?
Answer: Robotics technology improves the functionality and natural movement of prosthetic limbs, enhancing the quality of life for individuals with limb loss and disabilities.
100. How do robots equipped with artificial intelligence algorithms assist in the development of autonomous vehicles and self-driving cars?
Answer: Robots powered by AI algorithms process sensor data, make real-time decisions, and navigate autonomously, contributing to the development and safety of self-driving vehicles.
101. What role do robots play in the field of deep-sea mining, and how do they extract valuable resources from the ocean floor?
Answer: Robots are essential for deep-sea mining, where they collect mineral resources from the ocean floor using specialized equipment and remotely operated vehicles (ROVs).
102. How can robots assist in the monitoring and conservation of endangered marine species and coral reefs, particularly in the face of climate change?
Answer: Robots equipped with sensors and underwater cameras can monitor marine life, detect illegal fishing activities, and contribute to the conservation of endangered species and coral reefs.
103. What is the potential for robotics in the field of personalized healthcare, including customized medication and treatment plans?
Answer: Robots can analyze patient data to create personalized medication dosages, treatment plans, and rehabilitation exercises, improving healthcare outcomes.
104. How do robots contribute to the field of quantum computing and quantum technology research, including the manipulation of qubits?
Answer: Robots enable precise manipulation of qubits and experimental setups in quantum research, advancing the development of quantum computing and technologies.
105. What ethical challenges arise in the use of robots for elderly care, including the potential for isolation and the need for emotional support?
Answer: Ethical concerns include potential isolation and the emotional well-being of elderly individuals in care settings. Ethical guidelines should prioritize social interaction and emotional support in elderly care robotics.
106. How can robots support sustainable forest management practices, including monitoring deforestation and conducting reforestation efforts?
Answer: Robots can plant trees, monitor forest health, and combat illegal logging, contributing to sustainable forest management and biodiversity conservation.

107. What is the role of robots in the field of disaster prediction and early warning systems, such as earthquake and tsunami monitoring?

Answer: Robots equipped with seismic sensors and data analysis capabilities aid in earthquake prediction and tsunami monitoring, providing valuable data for early warning systems and disaster preparedness.

108. How do robots assist in the restoration of degraded ecosystems and reforestation efforts?

Answer: Robots can plant trees, restore damaged landscapes, and combat deforestation, aiding in the restoration of ecosystems and addressing climate change.

109. What is the potential for robotics in the development of sustainable and autonomous farming on Mars and other extraterrestrial environments?

Answer: Robots are critical for cultivating crops and managing resources in extraterrestrial farming, supporting future human colonization and resource utilization.

110. How can robotics and AI contribute to the development of autonomous marine vessels for scientific research and ocean exploration?

Answer: Robotics and AI enable autonomous marine vessels to conduct scientific research, map ocean depths, and monitor marine life, advancing our understanding of the world's oceans.

111. What ethical concerns surround the use of robots in journalism and media, particularly in generating news content and deepfake prevention?

Answer: Ethical dilemmas include misinformation, privacy violations, and deep fake creation. Responsible journalism ethics and technology safeguards are essential in robot-generated news content.

112. How can robots be employed in the field of art therapy and creative expression for individuals with mental health challenges?

Answer: Robots can assist in art therapy by providing creative tools, emotional support, and a non-judgmental environment, aiding individuals in managing mental health challenges.

113. What is the potential for robotics in the development of autonomous underground transportation systems, such as tunnel boring and subterranean transportation networks?

Answer: Robots play a crucial role in tunnel boring and maintenance for underground transportation systems, enabling efficient and safe subterranean travel and infrastructure development.

114. How do robots assist in the exploration and study of extreme environments on Earth, like active volcanoes and deep-sea hydrothermal vents?

Answer: Robots equipped with specialized sensors and materials can collect data and samples from extreme environments, enhancing our understanding of these unique ecosystems.

115. What ethical concerns arise in the use of social robots in eldercare and long-term care facilities, particularly regarding the emotional well-being of residents?

Answer: Ethical concerns include loneliness, attachment, and the potential for neglect in eldercare settings using social robots. Ethical guidelines should prioritize residents' emotional well-being and quality of life.

116. How can robots support the conservation of endangered species through assisted reproduction and breeding programs?

Answer: Robots can assist in assisted reproduction techniques for endangered species, such as artificial insemination and embryo transfer, helping to increase genetic diversity and prevent extinction.

117. What is the role of robotics in the development of autonomous underwater habitats and research stations for deep-sea exploration?

Answer: Robots contribute to the construction, maintenance, and operation of autonomous underwater habitats, supporting scientific research and human exploration of deep-sea environments.

118. How do robots assist in the management and maintenance of large-scale agricultural operations, such as agribusinesses and plantations?

Answer: Robots automate tasks in large-scale agriculture, such as planting, harvesting, and irrigation, optimizing resource use and crop production for agribusinesses.

119. What ethical considerations surround the use of robots for child education and companionship, particularly in terms of screen time and emotional development?

Answer: Ethical concerns include screen addiction and the impact on emotional development. Balancing screen time with real-world interactions is crucial in child education robotics.

120. How can robots contribute to the field of rehabilitation and therapy for individuals with substance abuse and addiction?

Answer: Robots can provide structured therapy sessions, motivational support, and monitoring for individuals in addiction recovery, complementing traditional rehabilitation programs.

121. What is the potential for robotics in the development of autonomous recycling and waste management systems for sustainable cities?

Answer: Robots can autonomously sort and process recyclable materials, reducing landfill waste and promoting sustainable waste management practices in urban areas.

122. How do robots with advanced artificial intelligence contribute to the development of autonomous vehicles for public transportation and smart city mobility solutions?

Answer: Robots with AI algorithms enable autonomous public transportation systems, reducing traffic congestion and emissions while improving urban mobility and transportation efficiency.

123. What is the role of robotics in the field of personalized medicine, including tailored drug delivery and treatment plans?

Answer: Robots assist in personalized medicine by delivering precise drug dosages, conducting targeted therapies, and tailoring treatment plans based on individual patient data.

124. How can robots support the preservation and restoration of cultural heritage sites, such as historical buildings and artworks?

Answer: Robots can perform delicate restoration tasks, assess structural integrity, and document cultural heritage sites without risking damage, aiding in their preservation and conservation.

125. What ethical challenges arise in the use of robots for military applications, including the potential for autonomous weapon systems and AI-driven warfare?

Answer: Ethical concerns include the loss of human control, civilian casualties, and the use of AI in warfare. International agreements and ethical principles are essential in addressing these challenges.

126. How do robots contribute to space tourism and the development of commercial space travel, including roles in spacecraft operation and maintenance?

Answer: Robots assist in spacecraft operation, maintenance, and ensuring passenger safety in the emerging field of space tourism, enabling commercial space travel for civilians.

127. What is the potential for robotics in the development of autonomous agricultural systems for sustainable food production on Earth and beyond?

Answer: Robotics technology can create autonomous agricultural systems for sustainable food production, addressing global food security challenges and supporting future space colonization.

128. How can robots be employed in the field of mental health and therapy, particularly in providing emotional support and therapeutic interventions?

Answer: Robots can offer emotional support, deliver therapeutic interventions, and assist mental health professionals in providing care and treatment for individuals with mental health issues.

129. What is the role of robots in the field of space debris removal and the mitigation of space debris hazards in Earth's orbit?

Answer: Robots equipped with capture mechanisms can remove and deorbit space debris, reducing the risk of collisions and hazards to operational satellites in Earth's orbit.

130. How do robots with advanced computer vision and machine learning algorithms contribute to the field of agriculture, including precision farming and pest control?

Answer: Robots with computer vision and machine learning can identify crop pests, monitor crop health, and optimize resource use in precision farming, improving agricultural efficiency and sustainability.

131. What is the significance of robotics in the development of autonomous lunar and Martian rovers for planetary exploration and research?

Answer: Robots play a crucial role in the exploration of celestial bodies like the Moon and Mars, assisting in data collection, sample analysis, and terrain mapping, which advances our understanding of these worlds and supports future human missions.

132. How can robots be employed in the field of biodiversity conservation, including the monitoring and protection of endangered species and ecosystems?

Answer: Robots with sensors and cameras can monitor wildlife, detect poaching activities, and collect environmental data, contributing to biodiversity conservation efforts and the protection of endangered species and ecosystems.

133. What ethical concerns arise in the use of robots for policing and law enforcement, particularly regarding surveillance, bias, and the use of force?

Answer: Ethical concerns include privacy infringements, potential biases in algorithms, and the use of force by autonomous robots. Ethical guidelines and transparency are essential to address these issues in policing.

134. How do robots assist in the development of innovative solutions for recycling and waste reduction, including recycling robots and waste-to-energy technologies?

Answer: Robots can sort and process recyclable materials more efficiently, contributing to waste reduction and the development of sustainable waste management solutions, such as waste-to-energy conversion.

135. What is the role of robotics in the field of regenerative medicine, including tissue engineering and organ transplantation?

Answer: Robots support regenerative medicine by automating tissue engineering processes, including the fabrication of artificial organs and tissues, facilitating advancements in regenerative therapies and organ transplantation.

136. How can robots contribute to the field of disaster resilience and preparedness, including the development of disaster-resistant infrastructure and early warning systems?

Answer: Robots can assist in the construction of disaster-resistant infrastructure, monitor disaster-prone areas, and provide data for early warning systems, enhancing disaster resilience and preparedness.

137. What ethical considerations surround the use of robots in journalism and content creation, especially in terms of misinformation and deep fake generation?

Answer: Ethical concerns include the potential for misinformation and deep fake creation. Adhering to ethical journalism standards and responsible content creation practices is crucial in robot-generated content.

138. How do robots assist in the maintenance and monitoring of remote and off-grid renewable energy installations, such as solar and wind farms?

Answer: Robots can inspect, clean, and maintain remote renewable energy installations, ensuring optimal energy generation and reducing the need for human interventions in challenging environments.

139. What is the potential for robotics in the field of personalized education, including adaptive learning and tailored curricula?

Answer: Robotics can support personalized education by providing tailored learning experiences, adapting to individual student needs, and offering interactive lessons to enhance engagement and learning outcomes.

140. How can robots aid in the conservation and protection of marine ecosystems, including coral reefs, seagrass beds, and marine sanctuaries?

Answer: Robots equipped with sensors and cameras can monitor marine ecosystems, assess coral health, and deter illegal fishing activities, contributing to the conservation and protection of marine environments.

141. What is the role of robots in the development of sustainable and autonomous transportation systems, including electric and self-driving vehicles for urban mobility?

Answer: Robots contribute to the development of autonomous and electric transportation systems by assisting in vehicle production, maintenance, and the implementation of smart city mobility solutions.

142. How do robots with advanced mobility systems assist in the exploration and study of extreme environments on other planets, such as Titan and Europa?

Answer: Robots equipped for extreme environments can explore and study the unique conditions on celestial bodies like Titan and Europa, uncovering insights into the potential for extraterrestrial life.

143. What ethical challenges arise in the use of robots for eldercare and assisted living, particularly in terms of privacy, autonomy, and informed consent?

Answer: Ethical concerns include privacy infringements, autonomy issues, and informed consent for elderly individuals. Ethical guidelines should prioritize individual rights and well-being in eldercare robotics.

144. How can robots support sustainable agriculture practices, including precision farming and organic farming methods?

Answer: Robots contribute to sustainable agriculture by optimizing resource use, monitoring soil health, and reducing the need for chemical inputs, supporting both precision and organic farming approaches.

145. What is the potential for robotics in the development of autonomous and efficient construction processes for infrastructure projects?

Answer: Robots can automate construction tasks, reducing labor costs and project duration, while improving safety and precision in infrastructure development.

146. How do robots assist in the development of innovative solutions for waste recycling, including plastic recycling and e-waste management?

Answer: Robots can sort and process recyclable materials, enhancing the efficiency of waste recycling, including plastic recycling and e-waste management, promoting a circular economy.

ONLINE RESOURCES

Practice Paper Link: https://forms.gle/kFNtniUvGCYadSKs9

QR Code:

Video Content Link:

https://www.youtube.com/playlist?list=PLzlh_31VWEFrlHrrc3nzJE3RaiVnLhaeo

QR Code:

Chapter 3: Robot Programming

1. What is robot programming?

Answer: Robot programming involves creating instructions that control a robot's movements and actions to perform specific tasks or operations.

2. What are the common programming languages used in robot programming?

Answer: Common languages include C++, Python, and ROS (Robot Operating System) for robotics-specific applications.

3. How do robots interpret programming instructions?

Answer: Robots rely on a controller or computer to interpret and execute programmed instructions.

4. What is the significance of kinematics in robot programming?

Answer: Kinematics helps calculate and control a robot's position and movement in three-dimensional space.

5. What are the key components of a robot's programming environment?

Answer: Components include a code editor, simulator, and communication tools for interfacing with the robot.

6. What is the difference between offline and online robot programming?

Answer: Offline programming involves writing and testing code on a computer, while online programming is done directly on the robot's controller.

7. How can inverse kinematics be applied in robot programming?

Answer: Inverse kinematics calculates joint angles to achieve a desired end-effector position, useful in tasks like pick-and-place operations.

8. What role does sensor integration play in robot programming?

Answer: Sensors provide real-time data that robots use to adapt their behavior or respond to the environment.

9. How does trajectory planning influence robot programming?

Answer: Trajectory planning determines the path and speed of a robot's movement to optimize efficiency and safety.

10. What are robot teach pendants, and how are they used in programming?

Answer: Teach pendants are handheld devices that allow operators to manually guide a robot and record positions, simplifying programming.

11. What is the purpose of motion planning in robot programming?

Answer: Motion planning algorithms find collision-free paths for robots, ensuring safe and efficient movements.

12. How can robots be programmed for autonomous navigation in dynamic environments?

Answer: Programming includes obstacle detection and avoidance algorithms, enabling robots to navigate autonomously.

13. What is ROS (Robot Operating System) and its role in robot programming?

Answer: ROS is an open-source framework that provides libraries and tools for developing and controlling robots, simplifying robot programming tasks.

14. What programming challenges are associated with multi-robot coordination and collaboration?

Answer: Challenges include communication protocols and coordination algorithms for synchronized actions.

15. What are the ethical considerations in robot programming, especially in autonomous systems?

Answer: Ethical issues involve decision-making algorithms, accountability, and potential consequences in situations involving human safety.

16. How can robots be programmed to interact with humans in a socially acceptable manner?

Answer: Programming includes natural language processing, gesture recognition, and social cues to facilitate human-robot interactions.

17. What is the role of machine learning and AI in advanced robot programming?

Answer: Machine learning and AI enable robots to adapt and learn from data, improving their decision-making and problem-solving capabilities.

18. How can robots be programmed to perform tasks in unstructured environments, like disaster response scenarios?

Answer: Programming involves adaptive algorithms that allow robots to respond to unpredictable conditions and obstacles.

19. What are the programming considerations for industrial robots used in manufacturing automation?

Answer: Considerations include precision, repeatability, and synchronization with other machines in the production line.

20. What programming techniques are used for swarm robotics applications involving multiple robots?

Answer: Techniques include distributed algorithms and coordination strategies to enable swarm behaviors.

21. What are the key factors to consider when programming robots for medical applications, such as surgery or patient care?

Answer: Factors include safety, accuracy, and the ability to work in close proximity to humans.

22. How can robots be programmed for tasks involving object recognition and manipulation?

Answer: Programming includes computer vision algorithms for object detection and grasping strategies for manipulation.

23. What is the role of simulation environments in robot programming?

Answer: Simulation environments allow developers to test and refine robot programs in a virtual setting before deployment.

24. How can robots be programmed to adapt to changing environmental conditions, such as weather or lighting variations?

Answer: Adaptive algorithms and sensor fusion techniques enable robots to respond to environmental changes effectively.

25. What is the significance of real-time control in robot programming, especially for applications requiring quick responses?

Answer: Real-time control ensures that robots can react to dynamic situations and make rapid adjustments as needed.

26. How do robots with swarm intelligence capabilities coordinate their actions and behaviors?

Answer: Robots communicate and share information through swarm intelligence algorithms to achieve coordinated actions.

27. What programming methodologies are used for humanoid robots to mimic human movements and behaviors?

Answer: Methods include motion capture, kinematic modeling, and imitation learning to replicate human-like actions.

28. What challenges are associated with programming robots for space exploration and extraterrestrial missions?

Answer: Challenges include communication delays, extreme environments, and autonomous decision-making in remote locations.

29. How can robots be programmed to assist individuals with disabilities in daily tasks and mobility?

Answer: Programming includes personalized adaptations to help individuals with disabilities perform specific tasks and enhance their mobility.

30. What are the implications of using robots in educational settings, and how can programming contribute to student learning?

Answer: Robots in education can enhance engagement and learning through interactive programming activities and hands-on experiences.

31. What is the role of robot programming in the development of autonomous delivery robots for logistics and e-commerce?

Answer: Robot programming is crucial for designing algorithms that allow delivery robots to navigate city streets, avoid obstacles, and make safe deliveries. These programs integrate mapping, localization, and path planning to optimize delivery routes and efficiency.

32. How do robots in the field of agriculture use programming to perform tasks like precision planting, weeding, and harvesting?

Answer: Agricultural robots are programmed to perform tasks like planting and harvesting with precision. They use computer vision and sensor data to identify crop conditions and adapt their actions accordingly, maximizing crop yield and minimizing resource usage.

33. What programming challenges are associated with robots used in underwater exploration and research, such as remotely operated vehicles (ROVs)?

Answer: Programming underwater ROVs involves addressing communication delays, maintaining stability, and navigating challenging underwater environments. Algorithms must account for buoyancy, water currents, and the need for real-time data transmission.

34. How can robots be programmed to work collaboratively with humans in manufacturing settings, ensuring safety and productivity?

Answer: Collaborative robot programming focuses on creating control algorithms that allow robots to work alongside humans safely. These robots use force sensors and vision systems to detect human presence and adjust their actions accordingly, facilitating productive cooperation.

35. What programming techniques enable swarm robotics systems to exhibit emergent behaviors, such as flocking or pattern formation?

Answer: Swarm robotics programming involves decentralized control algorithms that allow robots to communicate and coordinate their actions. These algorithms enable emergent behaviors like flocking, where robots move together in a coordinated manner.

36. How do robots used in the field of telemedicine employ programming to assist healthcare professionals in remote consultations and surgeries?

Answer: Telemedicine robots use programming to provide remote access to healthcare professionals. These robots can facilitate video consultations, deliver medications, or assist in surgical procedures with the guidance of a remote surgeon, enhancing healthcare access.

37. What programming considerations are essential for humanoid robots used in research on human-robot interaction and social robotics?

Answer: Programming humanoid robots for social interaction requires natural language processing, gesture recognition, and emotion recognition algorithms. These robots must exhibit socially acceptable behavior, including maintaining eye contact and responding appropriately to human cues.

38. What programming techniques are used in the development of autonomous drones for applications like aerial surveillance and package delivery?

Answer: Autonomous drones employ programming techniques such as path planning, obstacle avoidance, and image recognition. These algorithms enable drones to navigate complex environments, capture images, and make deliveries with precision.

39. How can robot programming contribute to the development of exoskeletons and wearable devices for rehabilitation and mobility assistance?

Answer: Robot programming plays a crucial role in designing control systems for exoskeletons and wearable devices. These programs enable users to regain mobility and assist in rehabilitation by interpreting user movements and providing support where needed.

40. What programming challenges are associated with the development of robots for space exploration and colonization on Mars or the Moon?

Answer: Space exploration robots require programming that accounts for communication delays, harsh environmental conditions, and the need for autonomy. These robots must operate in extreme environments, adapt to changing terrain, and make critical decisions independently.

41. What is the significance of robot programming in the field of autonomous drones used for environmental monitoring and wildlife conservation?

Answer: Robot programming is essential in equipping autonomous drones with flight control algorithms, data collection routines, and obstacle avoidance systems. These programs enable drones to efficiently monitor ecosystems, gather environmental data, and protect wildlife without human intervention.

42. How do robots programmed with natural language understanding contribute to language translation and communication assistance in diverse global settings?
Answer: Robots programmed with natural language understanding algorithms can provide real-time language translation services, facilitating communication between individuals who speak different languages. This technology is valuable in multicultural environments, international business, and global diplomacy.
43. What programming techniques are employed in swarm robotics for applications like search and rescue missions and environmental monitoring?
Answer: Swarm robotics programming involves coordination algorithms that enable robots to work together in teams. These algorithms take into account communication, sensing, and decision-making, making swarm robotics effective in scenarios such as search and rescue operations and environmental data collection.
44. How can robot programming support the development of autonomous electric vehicles for public transportation and smart city mobility solutions?
Answer: Robot programming is crucial for autonomous electric vehicles, enabling them to navigate city streets, interpret traffic signals, and interact with pedestrians. These programs ensure the safety and efficiency of public transportation and contribute to smart city mobility solutions.
45. What are the programming considerations for robots used in the agricultural sector to perform tasks like fruit picking and crop spraying?
Answer: Programming agricultural robots involves creating algorithms that use computer vision and machine learning to identify ripe fruit or assess crop health. These robots must coordinate movements with precision, optimizing resource usage and crop yield.
46. How do robot programming techniques enable the development of robotic exposits for enhancing human performance and mobility?
Answer: Robot programming for exposits focuses on designing control algorithms that assist users in tasks like walking, lifting, and carrying heavy loads. These programs use sensors to detect user movements and provide the necessary support, improving mobility and reducing physical strain.
47. What programming challenges arise in the deployment of swarm robotics systems for tasks like environmental cleanup and disaster response?
Answer: Programming swarm robotics systems for environmental cleanup and disaster response involves developing algorithms for collaborative navigation, task allocation, and communication. These programs must adapt to dynamic environments and unexpected obstacles while ensuring efficient coordination among robots.
48. How can robots be programmed to assist in archaeological excavations, including artifact identification and documentation?
Answer: Robot programming for archaeology involves creating algorithms that enable robots to navigate excavation sites, detect artifacts, and document findings. These programs are critical for preserving historical treasures and advancing archaeological research.
49. What is the role of robot programming in the development of autonomous submersibles for deep-sea exploration and the study of underwater ecosystems?
Answer: Robot programming is essential for autonomous submersibles used in deep-sea exploration. These programs control navigation, data collection, and communication, allowing submersibles to operate in challenging underwater environments and study marine life and ecosystems.
50. How can robot programming contribute to the field of personalized healthcare, including the development of robotic assistants for patient monitoring and rehabilitation?
Answer: Robot programming plays a significant role in the development of personalized healthcare robots. These programs enable robots to collect patient data, deliver tailored rehabilitation exercises, and provide emotional support, enhancing the quality of care and patient outcomes.
51. What programming techniques are employed in the creation of autonomous cleaning robots for tasks like vacuuming and floor mopping?

Answer: Programming autonomous cleaning robots involves algorithms for mapping and navigation. These programs help robots efficiently cover cleaning areas, avoid obstacles, and adapt to different floor surfaces, improving cleaning performance.

52. What programming considerations are essential for the development of robots used in the film and entertainment industry for special effects and character animation?

Answer: Programming robots for the film and entertainment industry involves creating control algorithms that synchronize movements with precise timing. These programs ensure that robots perform actions in coordination with actors and scene requirements, enhancing visual effects and character animations.

53. How do robot programming techniques enable the development of robotic prosthetics and assistive devices for individuals with limb loss and disabilities?

Answer: Robot programming for prosthetics and assistive devices focuses on creating control systems that respond to user input and provide natural and intuitive movements. These programs enhance the mobility and quality of life for individuals with limb loss and disabilities.

54. What are the programming challenges associated with robots used in the mining industry for tasks like drilling, excavation, and material transport?

Answer: Programming mining robots involves addressing challenges related to navigating complex underground environments, coordinating drilling and excavation tasks, and optimizing material transport routes. These programs ensure safe and efficient mining operations.

55. How can robots be programmed to support the exploration and study of extreme environments on Earth, such as active volcanoes and deep-sea hydrothermal vents?

Answer: Programming robots for extreme environments involves algorithms for sensor integration, heat resistance, and remote operation. These programs enable robots to collect data and samples from extreme environments while withstanding harsh conditions.

56. What programming techniques are employed in the development of autonomous agricultural machinery for tasks like plowing, planting, and harvesting?

Answer: Programming autonomous agricultural machinery involves creating control algorithms that automate farming tasks. These programs use sensor data to determine optimal planting depths, harvesting times, and resource management, enhancing agricultural efficiency.

57. How do robot programming methods contribute to the development of autonomous flying taxis and urban air mobility solutions?

Answer: Robot programming for flying taxis involves algorithms for flight control, obstacle avoidance, and navigation in urban airspace. These programs enable safe and efficient transportation within densely populated cities.

58. What is the role of robot programming in the field of remote sensing and environmental monitoring, including satellite and drone-based data collection?

Answer: Robot programming is essential for controlling satellites and drones used in remote sensing and environmental monitoring. These programs dictate flight paths, sensor operation, and data transmission, enabling the collection of valuable Earth observation data.

59. How can robots be programmed to assist in archaeological excavations, including the identification and preservation of fragile historical artifacts?

Answer: Robot programming for archaeology involves creating algorithms that enable robots to excavate with precision, detect artifacts, and handle fragile items. These programs ensure the careful preservation and documentation of historical relics.

60. What programming challenges arise in the deployment of swarm robotics systems for tasks like agricultural pollination and crop protection?

Answer: Programming swarm robotics systems for agricultural applications involves developing algorithms for coordinated pollination and pest detection. These programs ensure that robots work together efficiently to enhance crop yield and protect plants.

61. What programming techniques are employed in the development of autonomous underwater vehicles (AUVs) for marine research and exploration?

Answer: Programming AUVs involves algorithms for underwater navigation, data collection, and autonomous decision-making. These programs enable AUVs to explore and study the ocean's depths, collect scientific data, and contribute to marine research.

62. How can robot programming contribute to the field of personalized nutrition and dietary guidance by creating robotic chefs and meal preparation assistants?

Answer: Robot programming for personalized nutrition involves creating algorithms that consider dietary preferences and nutritional requirements. These programs enable robotic chefs to prepare customized meals, ensuring individuals receive tailored dietary guidance and healthy food choices.

63. What programming considerations are essential for robots used in the aerospace industry for tasks like aircraft maintenance and inspection?

Answer: Programming aerospace robots involves developing algorithms for precision inspections, maintenance tasks, and quality control. These programs ensure the safety and efficiency of aircraft maintenance operations.

64. How do robot programming methods enable the development of autonomous underwater habitats for extended deep-sea exploration and research?

Answer: Robot programming for underwater habitats involves algorithms for life support systems, environmental monitoring, and communication with the surface. These programs support prolonged deep-sea exploration and scientific research within underwater habitats.

65. What are the programming challenges associated with the development of robots used in the construction industry for tasks like bricklaying and concrete pouring?

Answer: Programming construction robots involves addressing challenges related to site navigation, material handling, and precision in tasks like bricklaying and concrete pouring. These programs ensure the accuracy and efficiency of construction processes.

66. How can robots be programmed to support the development of renewable energy solutions, including solar panel installation and maintenance?

Answer: Robot programming plays a crucial role in automating tasks related to renewable energy, such as the installation and maintenance of solar panels. These programs ensure the efficiency and longevity of renewable energy systems.

67. What programming techniques are employed in the creation of robotic companions for elderly individuals, including tasks like medication reminders and emotional support?

Answer: Programming robotic companions for the elderly involves algorithms for natural language understanding, emotional recognition, and healthcare reminders. These programs enhance the well-being and companionship of elderly individuals.

68. How do robot programming methods enable the development of autonomous underwater vehicles (AUVs) for marine research and exploration?

Answer: Programming AUVs involves algorithms for underwater navigation, data collection, and autonomous decision-making. These programs enable AUVs to explore and study the ocean's depths, collect scientific data, and contribute to marine research.

69. What are the programming considerations for robots used in the aerospace industry for tasks like aircraft maintenance and inspection?

Answer: Programming aerospace robots involves developing algorithms for precision inspections, maintenance tasks, and quality control. These programs ensure the safety and efficiency of aircraft maintenance operations.

70. How can robot programming support the development of renewable energy solutions, including solar panel installation and maintenance?

Answer: Robot programming plays a crucial role in automating tasks related to renewable energy, such as the installation and maintenance of solar panels. These programs ensure the efficiency and longevity of renewable energy systems.

71. What programming techniques are employed in the creation of robotic companions for elderly individuals, including tasks like medication reminders and emotional support?

Answer: Programming robotic companions for the elderly involves algorithms for natural language understanding, emotional recognition, and healthcare reminders. These programs enhance the well-being and companionship of elderly individuals.

72. How can robot programming contribute to the development of autonomous cargo ships and maritime transportation solutions?

Answer: Robot programming is integral in designing control systems for autonomous cargo ships. These programs manage navigation, collision avoidance, and cargo handling, optimizing maritime transportation for efficiency and safety.

73. What programming challenges arise in the deployment of swarm robotics systems for tasks like environmental monitoring and wildlife tracking in remote ecosystems?

Answer: Programming swarm robots for remote environmental monitoring requires algorithms for communication, coordination, and energy management. These programs enable efficient data collection and tracking in challenging and remote environments.

74. How do robot programming methods enable the development of autonomous drones for applications like precision agriculture and crop monitoring?

Answer: Drone programming involves algorithms for autonomous flight, data collection, and image analysis. These programs empower drones to gather crucial agricultural data, such as crop health and resource management, enhancing farming practices.

75. What programming considerations are essential for robots used in healthcare settings to perform tasks like patient lifting and assisting medical professionals?

Answer: Healthcare robot programming prioritizes safety, precision, and patient comfort. These programs ensure that robots can assist medical professionals in tasks like patient lifting, allowing for efficient and injury-free care.

76. How can robots be programmed to support underwater archaeology, including tasks like artifact retrieval and site documentation?

Answer: Robot programming for underwater archaeology focuses on creating control algorithms for robotic arms and manipulators. These programs allow robots to retrieve artifacts and document underwater archaeological sites with precision and care.

77. What programming techniques are employed in the development of autonomous inspection robots for tasks like pipeline monitoring and maintenance in the oil and gas industry?

Answer: Programming inspection robots in the oil and gas industry involves algorithms for navigation through complex environments, leak detection, and corrosion assessment. These programs ensure the safety and reliability of critical infrastructure.

78. How do robot programming methods enable the development of autonomous mining equipment for excavation and resource extraction?

Answer: Programming mining robots includes algorithms for excavation, resource detection, and haulage. These programs ensure the efficiency and safety of mining operations while minimizing human exposure to hazardous conditions.

79. What are the programming challenges associated with the development of robots used in disaster recovery missions, such as earthquake response and rubble removal?

Answer: Disaster recovery robots require programming for navigation in unstable environments, debris clearing, and victim detection. These programs enable robots to assist in life-saving efforts and efficient disaster response.

80. How can robot programming contribute to the field of precision medicine by assisting in the analysis of genetic data and personalized treatment recommendations?

Answer: Robot programming in precision medicine involves algorithms for genetic data analysis and treatment recommendations. These programs assist healthcare professionals in tailoring treatment plans to individual patients, improving healthcare outcomes.

81. What programming techniques are employed in the creation of autonomous drones for applications like wildlife conservation and anti-poaching efforts in remote areas?
Answer: Drone programming for wildlife conservation includes algorithms for long-range flight, wildlife tracking, and anti-poaching surveillance. These programs support the protection of endangered species and ecosystems in remote regions.
82. How do robot programming methods enable the development of autonomous underwater vehicles (AUVs) for scientific research and data collection in oceanography?
Answer: AUV programming includes algorithms for precise navigation, data collection, and environmental monitoring. These programs enable AUVs to explore and study oceanographic phenomena, contributing to scientific research.
83. What programming considerations are essential for robots used in the construction industry to perform tasks like concrete pouring and structural assembly?
Answer: Construction robot programming involves algorithms for precise task execution, structural assessment, and safety monitoring. These programs ensure the accuracy and efficiency of construction processes.
84. How can robots be programmed to support the agricultural industry in tasks like soil analysis and irrigation management?
Answer: Agricultural robots use programming to conduct soil analysis, monitor crop health, and manage irrigation. These programs optimize resource allocation and enhance crop yield while conserving water.
85. What programming techniques are employed in the development of autonomous underwater robots for the exploration and mapping of underwater caves and uncharted regions?
Answer: Programming underwater robots for cave exploration includes algorithms for navigation in confined spaces, obstacle avoidance, and 3D mapping. These programs enable the safe and thorough exploration of underwater caves.
86. How do robot programming methods enable the development of autonomous inspection robots for tasks like bridge and infrastructure assessment in civil engineering?
Answer: Inspection robot programming involves algorithms for precise traversal, defect detection, and structural assessment. These programs ensure the safety and integrity of civil engineering infrastructure.
87. What are the programming challenges associated with the development of robots used in agriculture for tasks like fruit picking and pesticide application?
Answer: Agriculture robots require programming for accurate fruit identification, precise harvesting, and pesticide application. These programs optimize farming practices and minimize environmental impact.
88. How can robot programming contribute to the field of space exploration by enabling autonomous rovers and landers for planetary missions?
Answer: Robot programming is crucial for designing control systems for autonomous rovers and landers. These programs enable navigation, sample collection, and data transmission, supporting planetary exploration and research.
89. What programming techniques are employed in the creation of autonomous drones for environmental monitoring in remote and ecologically sensitive areas?
Answer: Drone programming for remote environmental monitoring includes algorithms for autonomous flight planning, data collection, and ecological analysis. These programs facilitate data-driven conservation efforts in remote and fragile ecosystems.
90. How do robot programming methods enable the development of autonomous underwater vehicles (AUVs) for the study of deep-sea ecosystems and hydrothermal vents?
Answer: AUV programming for deep-sea exploration involves algorithms for navigation, data collection, and biological sampling. These programs enable AUVs to study and document deep-sea ecosystems, contributing to scientific understanding.
91. What programming considerations are essential for robots used in the food industry to perform tasks like food sorting and quality control?

Answer: Programming for food industry robots involves algorithms for food identification, sorting, and quality assessment. These programs enhance food safety and quality assurance.

92. How can robots be programmed to support archaeological research by assisting in the excavation of historical sites and ancient artifacts?

Answer: Robot programming for archaeology includes control algorithms for precise excavation and artifact handling. These programs ensure the careful preservation and documentation of historical discoveries.

93. What programming techniques are employed in the development of autonomous drones for disaster relief efforts, including search and rescue missions and aerial assessment?

Answer: Drone programming for disaster relief includes algorithms for search and rescue coordination, aerial mapping, and victim detection. These programs support rapid response and efficient disaster management.

94. How do robot programming methods enable the development of autonomous inspection robots for tasks like bridge and infrastructure assessment in civil engineering?

Answer: Inspection robot programming involves algorithms for precise traversal, defect detection, and structural assessment. These programs ensure the safety and integrity of civil engineering infrastructure.

95. What are the programming challenges associated with the development of robots used in agriculture for tasks like fruit picking and pesticide application?

Answer: Agriculture robots require programming for accurate fruit identification, precise harvesting, and pesticide application. These programs optimize farming practices and minimize environmental impact.

96. How can robot programming contribute to the field of space exploration by enabling autonomous rovers and landers for planetary missions?

Answer: Robot programming is crucial for designing control systems for autonomous rovers and landers. These programs enable navigation, sample collection, and data transmission, supporting planetary exploration and research.

97. What programming techniques are employed in the creation of autonomous drones for environmental monitoring in remote and ecologically sensitive areas?

Answer: Drone programming for remote environmental monitoring includes algorithms for autonomous flight planning, data collection, and ecological analysis. These programs facilitate data-driven conservation efforts in remote and fragile ecosystems.

98. How do robot programming methods enable the development of autonomous underwater vehicles (AUVs) for the study of deep-sea ecosystems and hydrothermal vents?

Answer: AUV programming for deep-sea exploration involves algorithms for navigation, data collection, and biological sampling. These programs enable AUVs to study and document deep-sea ecosystems, contributing to scientific understanding.

99. What programming considerations are essential for robots used in the food industry to perform tasks like food sorting and quality control?

Answer: Programming for food industry robots involves algorithms for food identification, sorting, and quality assessment. These programs enhance food safety and quality assurance.

100. How can robots be programmed to support archaeological research by assisting in the excavation of historical sites and ancient artifacts?

Answer: Robot programming for archaeology includes control algorithms for precise excavation and artifact handling. These programs ensure the careful preservation and documentation of historical discoveries.

101. What is the role of robot programming in the development of autonomous vehicles for long-haul trucking and freight transportation?

Answer: Robot programming is crucial in creating control systems for autonomous trucks. These programs manage route planning, obstacle detection, and cargo management, enhancing efficiency and safety in freight transportation.

102. How do robot programming methods enable the development of autonomous drones for applications like disaster relief, including the delivery of medical supplies and humanitarian aid?

Answer: Drone programming for disaster relief involves algorithms for flight coordination, cargo delivery, and emergency response planning. These programs support rapid aid delivery and lifesaving efforts in disaster-stricken areas.

103. What programming challenges arise in the deployment of swarm robotics systems for tasks like environmental monitoring in remote and ecologically diverse ecosystems?

Answer: Programming swarm robots for remote environmental monitoring requires algorithms for adaptive decision-making, communication, and resource management. These programs facilitate efficient data collection and monitoring in diverse and challenging environments.

104. How can robot programming contribute to the field of precision agriculture by enabling autonomous tractors and farm machinery for crop cultivation?

Answer: Robot programming plays a significant role in designing control systems for autonomous tractors and farm equipment. These programs manage tasks such as plowing, planting, and irrigation, optimizing crop yield and resource usage.

105. What programming techniques are employed in the development of autonomous drones for applications like wildlife conservation and animal tracking in remote regions?

Answer: Drone programming for wildlife conservation includes algorithms for wildlife tracking, habitat assessment, and anti-poaching efforts. These programs aid in the protection of endangered species and ecosystems in remote areas.

106. How do robot programming methods enable the development of autonomous underwater vehicles (AUVs) for marine archaeology and the exploration of shipwrecks?

Answer: AUV programming for marine archaeology involves algorithms for precise underwater navigation, artifact documentation, and site mapping. These programs support the exploration and preservation of underwater cultural heritage.

107. What programming considerations are essential for robots used in precision manufacturing to perform tasks like microelectronics assembly and quality control?

Answer: Programming for precision manufacturing robots includes algorithms for precision assembly, defect detection, and quality assessment. These programs ensure high-precision manufacturing processes and product quality.

108. How can robots be programmed to support the construction industry by performing tasks like bricklaying, concrete reinforcement, and structural welding?

Answer: Robot programming in construction involves control algorithms for precise task execution, structural reinforcement, and safety monitoring. These programs enhance construction efficiency and structural integrity.

109. What programming techniques are employed in the development of autonomous drones for geological exploration and mineral resource assessment in remote areas?

Answer: Drone programming for geological exploration includes algorithms for terrain mapping, mineral detection, and geological surveying. These programs facilitate resource assessment and exploration in challenging and remote terrains.

110. How do robot programming methods enable the development of autonomous inspection robots for tasks like pipeline and underwater infrastructure maintenance?

Answer: Inspection robot programming involves algorithms for navigation in confined spaces, defect identification, and maintenance tasks. These programs ensure the safety and reliability of critical infrastructure.

111. What are the programming challenges associated with the development of robots used in the agriculture sector for tasks like crop spraying and soil analysis?

Answer: Agriculture robots require programming for precise soil analysis, targeted pesticide application, and resource management. These programs optimize farming practices, reduce environmental impact, and improve crop health.

112. How can robot programming contribute to the field of space exploration by enabling autonomous spacecraft for interplanetary missions and scientific research?

Answer: Robot programming is vital for designing control systems for autonomous spacecraft. These programs manage navigation, data collection, and communication, supporting interplanetary exploration and scientific research.

113. What programming techniques are employed in the creation of autonomous drones for wildfire monitoring, including early detection and real-time fire tracking?

Answer: Drone programming for wildfire monitoring includes algorithms for early fire detection, rapid response coordination, and fire tracking. These programs aid in wildfire prevention and containment efforts.

114. How do robot programming methods enable the development of autonomous underwater vehicles (AUVs) for marine biology research and the study of marine ecosystems?

Answer: AUV programming for marine biology involves algorithms for precise navigation, underwater data collection, and marine species tracking. These programs support scientific research and conservation efforts in marine ecosystems.

115. What programming considerations are essential for robots used in the pharmaceutical industry for tasks like drug manufacturing and laboratory automation?

Answer: Pharmaceutical robot programming includes algorithms for precision drug manufacturing, laboratory automation, and quality control. These programs ensure the accuracy and efficiency of pharmaceutical processes.

116. How can robots be programmed to assist in archaeological research by performing tasks like 3D scanning of historical artifacts and excavation site documentation?

Answer: Robot programming for archaeology includes control algorithms for 3D scanning, artifact documentation, and excavation site mapping. These programs enhance archaeological research and artifact preservation.

117. What programming techniques are employed in the development of autonomous drones for applications like environmental monitoring and pollution assessment in urban areas?

Answer: Drone programming for urban environmental monitoring includes algorithms for pollution detection, urban planning, and data analysis. These programs contribute to sustainable urban development and environmental protection.

118. How do robot programming methods enable the development of autonomous inspection robots for tasks like bridge and tunnel assessment in civil engineering?

Answer: Inspection robot programming involves algorithms for precise traversal, structural assessment, and defect identification. These programs ensure the safety and maintenance of critical civil engineering infrastructure.

119. What are the programming challenges associated with the development of robots used in the healthcare sector for tasks like surgery assistance and patient care?

Answer: Healthcare robots require programming for surgical precision, patient care assistance, and medical data management. These programs prioritize patient safety and well-being in medical settings.

120. How can robot programming contribute to the field of renewable energy by enabling autonomous maintenance robots for wind turbines and solar farms?

Answer: Robot programming is essential for designing control systems for autonomous maintenance robots in the renewable energy sector. These programs manage tasks like turbine inspection, cleaning, and component replacement, ensuring the efficiency and longevity of renewable energy systems.

121. What programming techniques are employed in the creation of autonomous drones for applications like archaeological site mapping and historical preservation?

Answer: Drone programming for archaeological mapping includes algorithms for site documentation, artifact identification, and historical preservation. These programs enhance the documentation and protection of cultural heritage sites.

122. How do robot programming methods enable the development of autonomous underwater vehicles (AUVs) for deep-sea exploration and the study of marine biodiversity?

Answer: AUV programming for deep-sea exploration involves algorithms for navigation in extreme depths, biological sampling, and biodiversity assessment. These programs advance our understanding of deep-sea ecosystems and marine life.

123. What programming considerations are essential for robots used in the aerospace industry for tasks like spacecraft assembly and component testing?

Answer: Aerospace robot programming includes algorithms for precision assembly, component testing, and quality control. These programs ensure the reliability and performance of aerospace systems.

124. How can robots be programmed to support agricultural research by performing tasks like plant phenotyping and crop disease detection?

Answer: Robot programming in agricultural research involves algorithms for plant phenotyping, disease detection, and data analysis. These programs enhance crop research and disease management.

125. What programming techniques are employed in the development of autonomous drones for applications like ecological monitoring and wildlife tracking in remote and protected areas?

Answer: Drone programming for ecological monitoring includes algorithms for wildlife tracking, habitat assessment, and environmental conservation efforts. These programs aid in the protection of sensitive ecosystems and species.

126. How do robot programming methods enable the development of autonomous inspection robots for tasks like railway track maintenance and monitoring in the transportation sector?

Answer: Inspection robot programming involves algorithms for railway track inspection, defect detection, and maintenance tasks. These programs ensure the safety and reliability of railway transportation.

127. What are the programming challenges associated with the development of robots used in the food industry for tasks like food packaging and quality assurance?

Answer: Programming for food industry robots includes algorithms for food packaging, quality assessment, and traceability. These programs ensure food safety and compliance with industry standards.

128. How can robot programming contribute to the field of marine science by enabling autonomous underwater vehicles (AUVs) for oceanographic research and data collection?

Answer: Robot programming is integral in designing control systems for AUVs in marine science. These programs manage navigation, data collection, and environmental monitoring, contributing to oceanographic research and data collection.

129. What programming techniques are employed in the creation of autonomous drones for applications like geological surveying and mineral exploration in remote and rugged terrains?

Answer: Drone programming for geological surveying includes algorithms for terrain mapping, mineral exploration, and geological data analysis. These programs support resource assessment and exploration in challenging terrains.

130. How do robot programming methods enable the development of autonomous inspection robots for tasks like underwater pipeline maintenance and offshore structure assessment in the energy sector?

Answer: Inspection robot programming involves algorithms for underwater navigation, structural assessment, and maintenance tasks in the energy sector. These programs ensure the safety and integrity of offshore structures and pipelines.

131. What is the role of robot programming in the development of autonomous agricultural robots for tasks like precision weeding and soil health analysis?

Answer: Robot programming is essential in designing control systems for autonomous agricultural robots. These programs enable precise weeding, soil health analysis, and resource management, enhancing crop yield and sustainability.

132. How can robot programming contribute to the field of underwater archaeology by enabling remotely operated vehicles (ROVs) for the exploration and documentation of submerged historical sites?

Answer: Robot programming for underwater archaeology involves control algorithms for ROVs. These programs facilitate the exploration, artifact documentation, and preservation of submerged historical sites, advancing underwater archaeology.

133. What programming techniques are employed in the development of autonomous drones for applications like atmospheric research and climate monitoring?

Answer: Drone programming for atmospheric research includes algorithms for data collection, weather pattern analysis, and climate monitoring. These programs contribute to our understanding of weather and climate dynamics.

134. How do robot programming methods enable the development of autonomous inspection robots for tasks like underwater pipeline inspection and maintenance in the oil and gas industry?

Answer: Inspection robot programming involves algorithms for underwater navigation, defect detection, and maintenance tasks in the oil and gas sector. These programs ensure the safety and reliability of underwater pipelines and infrastructure.

135. What are the programming challenges associated with the development of robots used in the automotive industry for tasks like car assembly and quality control?

Answer: Automotive robot programming includes algorithms for precision assembly, quality inspection, and production line coordination. These programs optimize automotive manufacturing processes and product quality.

136. How can robots be programmed to support the mining industry by performing tasks like mineral extraction and underground mapping in complex terrains?

Answer: Robot programming in mining involves algorithms for mineral extraction, underground mapping, and resource assessment. These programs enhance mining operations in challenging terrains and ensure resource efficiency.

137. What programming techniques are employed in the creation of autonomous drones for applications like disaster response and the delivery of humanitarian aid in remote and disaster-affected areas?

Answer: Drone programming for disaster response includes algorithms for rapid deployment, cargo delivery, and emergency response coordination. These programs aid in disaster relief efforts and the delivery of humanitarian aid.

138. How do robot programming methods enable the development of autonomous underwater vehicles (AUVs) for marine geology research and the study of geological features on the ocean floor?

Answer: AUV programming for marine geology involves algorithms for precise navigation, geological sampling, and seafloor mapping. These programs support scientific research and geological exploration in the marine environment.

139. What programming considerations are essential for robots used in the pharmaceutical industry for tasks like drug formulation and laboratory automation?

Answer: Pharmaceutical robot programming includes algorithms for precise drug formulation, laboratory automation, and quality control. These programs ensure the accuracy and efficiency of pharmaceutical processes and research.

140. How can robot programming contribute to the field of environmental protection by enabling autonomous drones for pollution monitoring and response in sensitive ecosystems?

Answer: Robot programming is vital in creating control systems for autonomous drones in environmental protection. These programs manage pollution monitoring, data analysis, and response efforts in sensitive ecosystems, contributing to environmental conservation.

141. What programming techniques are employed in the development of autonomous inspection robots for tasks like bridge and dam assessment in civil engineering?

Answer: Inspection robot programming includes algorithms for precise traversal, structural assessment, and defect detection in civil engineering infrastructure. These programs ensure the safety and maintenance of critical structures.

142. How do robot programming methods enable the development of autonomous robots for disaster recovery missions, including debris removal and building collapse search and rescue?

Answer: Robot programming for disaster recovery involves algorithms for navigation in hazardous environments, debris clearing, and victim detection. These programs support efficient disaster response and recovery efforts.

143. What are the programming challenges associated with the development of robots used in the aerospace industry for tasks like aircraft maintenance and engine inspection?

Answer: Aerospace robot programming includes algorithms for precision maintenance, engine inspection, and quality control in the aerospace sector. These programs ensure the reliability and safety of aircraft.

144. How can robots be programmed to support the agriculture sector by performing tasks like pest control and crop monitoring using advanced sensing technologies?

Answer: Robot programming in agriculture includes algorithms for pest control, crop monitoring, and sensor integration. These programs optimize pest management and enhance crop health using advanced sensing technologies.

145. What programming techniques are employed in the development of autonomous drones for applications like forest fire detection and firefighting support?

Answer: Drone programming for forest fire detection includes algorithms for early fire identification, firefighting coordination, and real-time monitoring. These programs aid in forest fire prevention and response efforts.

146. How do robot programming methods enable the development of autonomous underwater vehicles (AUVs) for marine pollution assessment and the study of oceanic waste disposal?

Answer: AUV programming for marine pollution assessment includes algorithms for navigation in polluted waters, data collection, and waste analysis. These programs support environmental monitoring and pollution management efforts.

147. What programming considerations are essential for robots used in the food processing industry for tasks like sorting and packaging of food products?

Answer: Programming for food processing robots includes algorithms for food sorting, packaging, and quality assurance. These programs ensure food safety, quality, and compliance with industry standards.

148. How can robot programming contribute to the field of marine science by enabling autonomous underwater vehicles (AUVs) for studying deep-sea ecosystems and hydrothermal vent communities?

Answer: Robot programming plays a crucial role in designing control systems for AUVs in marine science. These programs manage navigation, data collection, and scientific research, advancing our understanding of deep-sea ecosystems.

149. What programming techniques are employed in the creation of autonomous drones for applications like archaeological site preservation and protection against looting?

Answer: Drone programming for archaeological site protection includes algorithms for site documentation, artifact identification, and security monitoring. These programs aid in preserving and safeguarding cultural heritage sites.

150. How do robot programming methods enable the development of autonomous inspection robots for tasks like railway track maintenance and monitoring in the transportation sector?

Answer: Inspection robot programming involves algorithms for railway track inspection, defect detection, and maintenance tasks in the transportation sector. These programs ensure the safety and reliability of railway transportation systems.

151. What is the role of robot programming in the development of autonomous ships for oceanographic research, data collection, and marine exploration?

Answer: Robot programming is instrumental in creating control systems for autonomous ships used in oceanographic research and marine exploration. These programs manage navigation, data collection, and environmental monitoring, enabling autonomous vessels to traverse vast oceanic regions while collecting valuable scientific data. Additionally, they facilitate the safe operation of these vessels, ensuring compliance with maritime regulations and the efficient execution of research missions.

152. How can robot programming contribute to the field of precision forestry by enabling autonomous logging and tree health assessment?

Answer: Robot programming is pivotal in designing control systems for autonomous logging equipment and tree health assessment robots in precision forestry. These programs manage tasks such as tree identification, precise logging, and forest health monitoring, optimizing timber harvesting operations while minimizing environmental impact.

153. What programming techniques are employed in the development of autonomous drones for applications like marine wildlife tracking and conservation efforts in coastal regions?

Answer: Drone programming for marine wildlife tracking includes algorithms for real-time tracking, habitat assessment, and conservation efforts in coastal ecosystems. These programs support the protection of marine life and ecosystems, especially in areas affected by human activities.

154. How do robot programming methods enable the development of autonomous inspection robots for tasks like offshore wind turbine maintenance and underwater cable inspections in the renewable energy sector?

Answer: Robot programming for offshore wind turbine maintenance and underwater cable inspections involves algorithms for navigation in challenging marine environments, defect detection, and maintenance tasks. These programs ensure the reliability and performance of renewable energy infrastructure while reducing human risk.

155. What are the programming challenges associated with the development of robots used in the mining industry for tasks like autonomous drilling and mineral extraction in underground mines?

Answer: Mining robots require programming for autonomous drilling, mineral extraction, and underground navigation. These programs must address the complex geological and safety challenges of underground mining operations while maximizing resource efficiency.

156. How can robot programming contribute to the aerospace industry by enabling autonomous spacecraft for planetary exploration and space research?

Answer: Robot programming plays a pivotal role in designing control systems for autonomous spacecraft used in planetary exploration and space research. These programs manage navigation, data collection, and communication with Earth, supporting scientific missions and interplanetary exploration.

ONLINE RESOURCES

Practice Paper Link: https://forms.gle/SVdmeWuqHjjETj9v9

QR Code:

Video Content Link:

https://www.youtube.com/playlist?list=PLzlh_31VWEFrqfoUEzv8pKuGGHXpSRLnO

QR Code:

Chapter 4: Kinematics and Dynamics

1. What is robot kinematics?

Answer: Robot kinematics deals with the study of a robot's position and motion without considering the forces or torques involved. It focuses on describing how a robot moves in its environment.

2. What are the primary types of robot motion in kinematics?

Answer: In kinematics, robots can have translational motion (linear movement) and rotational motion (angular movement) or a combination of both.

3. How is forward kinematics used in robotics?

Answer: Forward kinematics calculates the end-effector's position and orientation based on the robot's joint angles or parameters. It helps determine where the robot's tool or end-effector is in space.

4. What is inverse kinematics, and why is it important in robotics?

Answer: Inverse kinematics finds the joint angles or parameters needed to achieve a desired end-effector position and orientation. It's crucial for planning and controlling a robot's movements accurately.

5. What are the degrees of freedom (DOF) of a robot, and why are they significant?

Answer: The DOF represent the number of independent ways a robot can move. Knowing a robot's DOF helps assess its mobility and ability to perform specific tasks.

6. How do you calculate the Jacobian matrix in robot kinematics?

Answer: The Jacobian matrix relates the change in joint variables to the change in end-effector position. It's computed by differentiating the forward kinematics equations with respect to joint variables.

7. What is robot dynamics in robotics?

Answer: Robot dynamics focuses on understanding the forces and torques acting on a robot and how they influence its motion. It deals with the acceleration, velocity, and forces involved in robotic systems.

8. Why is knowledge of robot dynamics essential for safe and efficient robotic operation?

Answer: Understanding robot dynamics is crucial for ensuring safe and efficient robot movements, as it helps calculate the forces and torques involved and prevents excessive or harmful motions.

9. What is the difference between forward and inverse dynamics in robotics?

Answer: Forward dynamics calculates the robot's motion based on given forces and torques, while inverse dynamics determines the required forces and torques to achieve a desired motion.

10. How does the concept of inertia affect robot dynamics?

Answer: Inertia represents an object's resistance to changes in motion. In robot dynamics, it accounts for the mass distribution and moments of inertia of robot links, affecting how forces and torques influence motion.

11. What is the significance of the Newton-Euler equations in robot dynamics?

Answer: The Newton-Euler equations are fundamental equations in robot dynamics, relating forces, torques, accelerations, and joint motions. They provide a framework for modeling and analyzing robot dynamics.

12. How are external forces and torques accounted for in robot dynamics?

Answer: External forces and torques, such as contact with objects or the environment, are considered in robot dynamics by including them as inputs to the equations of motion, helping predict robot behavior during interactions.

13. How does the concept of gravity compensation impact robot dynamics?

Answer: Gravity compensation is essential in robot dynamics to counteract the effects of gravity on the robot's joints and links, allowing for more precise control of the robot's motion.

14. What role does dynamic modeling play in robotics applications such as robot control and trajectory planning?

Answer: Dynamic modeling provides essential insights into how robots respond to various inputs, making it valuable for robot control, trajectory planning, and optimization of robotic tasks.

15. What are the limitations of rigid body dynamics in robotics, and when might flexible body dynamics be considered?

Answer: Rigid body dynamics assume that all robot links are rigid, which not hold may true in some cases. Flexible body dynamics considers deformations and is relevant when dealing with highly flexible robot components.

16. How can robot dynamics be used to analyze and improve the efficiency of robotic manipulators?
Answer: Robot dynamics analysis helps identify inefficient movements and forces, allowing for optimization of control strategies and design improvements to enhance robotic manipulator efficiency.
17. How does robot dynamics impact the development of robotic exoskeletons for human augmentation?
Answer: Robot dynamics play a critical role in designing exoskeletons that provide assistance and enhance human capabilities while ensuring natural and safe movements.
18. What challenges are associated with real-time computation of robot dynamics in applications like industrial automation?
Answer: Real-time computation of robot dynamics requires efficient algorithms and hardware, as it can be computationally intensive. Delays or inaccuracies in real-time calculations can impact robotic performance.
19. How does the concept of dynamic singularities affect robot dynamics, and how can they be avoided or mitigated?
Answer: Dynamic singularities are configurations where a robot experiences unusual dynamics. Careful planning and control strategies can help avoid or mitigate these situations to ensure smooth robot operation.
20. What advancements in computational methods have improved the accuracy and efficiency of robot dynamics simulations?
Answer: Advances in numerical methods, high-performance computing, and simulation software have significantly improved the accuracy and efficiency of robot dynamics simulations, allowing for more complex and realistic models.
21. What is the role of sensor feedback in controlling robot dynamics during tasks like walking for humanoid robots?
Answer: Sensor feedback, such as from accelerometers and force sensors, provides real-time information about the robot's state and environment. It is crucial for adjusting control signals to maintain stability and adapt to changing conditions during tasks like walking.
22. How does robot dynamics contribute to the development of robotic systems for space exploration and planetary rovers?
Answer: Understanding robot dynamics is essential for designing robotic systems that can navigate challenging environments like celestial bodies. It ensures that these systems can adapt to low-gravity conditions and operate effectively in space.
23. What is the impact of robot dynamics on the field of medical robotics, especially in surgical applications?
Answer: Robot dynamics in medical robotics plays a critical role in ensuring precise and stable movements during surgical procedures, minimizing the risk of errors and enhancing the surgeon's capabilities.
24. How can robot dynamics be used to improve the performance of mobile robots in scenarios like search and rescue missions?
Answer: Robot dynamics analysis helps optimize the design and control of mobile robots for search and rescue missions, allowing them to navigate complex terrains and overcome obstacles more effectively.
25. How does robot dynamics influence the development of legged robots for applications like exploration and locomotion in uneven terrain?
Answer: Robot dynamics considerations are central to designing legged robots that can walk, run, or traverse challenging terrains. It ensures the stability and agility needed for these robots to perform effectively in uneven environments.
26. What is the role of simulation software in studying and experimenting with robot dynamics without physical prototypes?
Answer: Simulation software enables engineers and researchers to model and analyze robot dynamics virtually, allowing for experimentation and testing of various scenarios without the need for physical prototypes, saving time and resources.

27. How can students and aspiring roboticists learn about robot kinematics and dynamics to build a strong foundation in robotics?

Answer: Students can study robot kinematics and dynamics through courses, textbooks, online resources, and practical hands-on projects. Developing a deep understanding of these concepts is essential for a successful career in robotics.

28. What emerging trends and technologies are shaping the field of robot kinematics and dynamics in robotics research and development?

Answer: Emerging trends include advanced control algorithms, machine learning for dynamic modeling, and the integration of sensors for real-time feedback, all contributing to more capable and adaptable robotic systems.

29. In what ways are researchers and engineers applying the principles of robot kinematics and dynamics to create innovative solutions in fields such as autonomous vehicles and drone technology?

Answer: The principles of robot kinematics and dynamics are crucial in developing control systems for autonomous vehicles and drones, ensuring precise and safe movements, collision avoidance, and efficient operation in various applications.

30. How can robot kinematics and dynamics be used to enhance human-robot collaboration and the development of collaborative robots (cobots)?

Answer: Understanding robot kinematics and dynamics is essential for designing collaborative robots that can work alongside humans safely and effectively. It enables cobots to adapt to human movements and tasks while maintaining safety and productivity.

31. How does the concept of singularities impact robot dynamics and kinematics?

Answer: Singularities in robot dynamics and kinematics are configurations where the robot's behavior becomes unpredictable or exhibits unusual characteristics. They can lead to reduced accuracy and control difficulties, making it important to identify and address them in robot design and control algorithms.

32. What role do joint constraints play in robot kinematics and dynamics, and how are they accounted for in robot control?

Answer: Joint constraints in robot kinematics and dynamics limit the range of motion for robot joints. These constraints must be considered in control algorithms to ensure that the robot operates within safe and allowable joint limits.

33. How are redundancy and over-constrained systems addressed in robot kinematics and dynamics?

Answer: Redundancy in robot kinematics refers to having more degrees of freedom than required for a task. Over-constrained systems can lead to kinematic issues. Techniques like optimization and null space control are used to address these challenges while optimizing task performance.

34. How can robot kinematics and dynamics be applied to humanoid robots to achieve natural and human-like movements?

Answer: Robot kinematics and dynamics are used to model and control humanoid robots, enabling them to imitate human movements. This includes walking, balancing, and interacting with the environment in a way that mimics human biomechanics.

35. What is the significance of dynamic modeling in the development of robotic prosthetics and exoskeletons for human mobility assistance?

Answer: Dynamic modeling is crucial in designing robotic prosthetics and exoskeletons that assist individuals with mobility impairments. It ensures that these devices provide natural, comfortable, and safe movements while interacting with the user's body and the surrounding environment.

36. How do robot kinematics and dynamics impact the field of virtual reality (VR) and augmented reality (AR) by enabling realistic simulations and interactions?

Answer: Robot kinematics and dynamics are essential in VR and AR applications, enabling realistic simulations of physical interactions. These technologies use dynamic modeling to create lifelike environments and user interactions, enhancing the immersive experience.

37. What is the role of robot kinematics and dynamics in haptic feedback systems for teleoperation and remote control of robotic devices?

Answer: Robot kinematics and dynamics are used to model the behavior of remote robotic devices and provide haptic feedback to operators. This feedback allows operators to feel and interact with the remote environment, improving control and situational awareness.

38. How can robot kinematics and dynamics contribute to the development of assistive devices for individuals with disabilities?

Answer: Robot kinematics and dynamics are applied to create assistive devices like robotic arms or exoskeletons, enabling individuals with disabilities to regain mobility and independence by enhancing their physical capabilities.

39. What is the role of optimization techniques in robot dynamics and kinematics, and how are they used to improve performance and efficiency?

Answer: Optimization techniques are applied to robot dynamics and kinematics to fine-tune control algorithms, minimize energy consumption, and optimize motion trajectories. These techniques help achieve more efficient and effective robotic movements.

40. How does the study of robot kinematics and dynamics impact the design and control of drones for aerial surveillance and delivery applications?

Answer: Understanding robot kinematics and dynamics is essential in designing control systems for drones used in surveillance and delivery. It ensures stable flight, efficient path planning, and effective payload delivery while accounting for external forces like wind.

41. In what ways do robot kinematics and dynamics influence the design of robotic arms used in manufacturing and assembly processes?

Answer: Robot kinematics and dynamics are critical for designing robotic arms that perform precise and efficient manufacturing and assembly tasks. They ensure accurate positioning and force control, optimizing industrial processes.

42. How can robot kinematics and dynamics be applied to underwater robots for tasks like ocean exploration and underwater infrastructure inspection?

Answer: Robot kinematics and dynamics are used in the development of underwater robots to model their motion and interactions with the underwater environment. This knowledge helps design robots that can operate effectively in underwater exploration and maintenance missions.

43. What challenges arise in controlling the dynamics of legged robots like quadrupeds and hexapods, and how are they overcome?

Answer: Legged robots face challenges in maintaining balance and stability due to their dynamic nature. These challenges are addressed through advanced control algorithms and adaptive strategies to ensure stable locomotion on various terrains.

44. How can robot kinematics and dynamics be applied to the design of robotic vehicles for planetary exploration and space missions?

Answer: Robot kinematics and dynamics are crucial for designing robotic vehicles that can navigate the unique challenges of celestial bodies, such as low gravity and uneven surfaces. These principles ensure the reliability and efficiency of space exploration missions.

45. What is the role of machine learning and artificial intelligence in enhancing robot kinematics and dynamics modeling and control?

Answer: Machine learning and AI techniques are used to improve the accuracy and adaptability of robot kinematics and dynamics models. They enable robots to learn from data, adapt to changing environments, and optimize their movements.

46. How do researchers leverage simulation-based approaches to study and experiment with robot kinematics and dynamics in complex scenarios?
Answer: Simulation-based approaches allow researchers to create virtual environments where robot kinematics and dynamics can be thoroughly tested and optimized before physical implementation. This reduces costs and speeds up development.
47. What impact do robot kinematics and dynamics have on the development of agile and versatile robots for tasks like disaster response and search-and-rescue missions?
Answer: Robot kinematics and dynamics influence the design of agile robots capable of navigating disaster-stricken areas and rough terrains. Understanding these principles helps optimize mobility and control in these critical applications.
48. How can robot kinematics and dynamics be integrated with computer vision to enable robots to perceive and respond to their surroundings more effectively?
Answer: Combining robot kinematics and dynamics with computer vision allows robots to interpret their environment, adapt to obstacles, and make informed decisions, enhancing their autonomy and situational awareness.
49. What role do sensor technologies, such as accelerometers and gyroscopes, play in improving robot dynamics modeling and control?
Answer: Sensors like accelerometers and gyroscopes provide real-time data about a robot's motion and orientation, enhancing dynamic modeling and control by enabling precise feedback and adjustment.
50. How does the concept of redundancy resolution in robot kinematics contribute to improving the reliability and versatility of robotic systems?
Answer: Redundancy resolution techniques allow robots to optimize their motions to achieve specific objectives while avoiding obstacles and joint limits. This enhances the adaptability and reliability of robotic systems in complex environments.
51. What are the implications of robot kinematics and dynamics on the development of soft robots with flexible and deformable structures?
Answer: Robot kinematics and dynamics for soft robots are complex due to their flexible nature. Understanding these aspects is crucial for designing control strategies that exploit the benefits of flexibility while ensuring stability and predictable motion.
52. How can robot kinematics and dynamics be used to create bio-inspired robots that mimic animal locomotion and behavior?
Answer: Robot kinematics and dynamics are applied to model and replicate animal movements in bio-inspired robots. This research field, known as biomimetic, aims to develop robots with natural and efficient locomotion for various applications.
53. What challenges arise in controlling the dynamics of aerial robots like quadcopters and fixed-wing drones, and how are they addressed?
Answer: Aerial robot dynamics are influenced by factors like wind and aerodynamics. Challenges in stability and control are addressed through advanced flight control algorithms and sensor integration.
54. What role does real-time feedback play in ensuring precise and responsive robot movements in dynamic environments?
Answer: Real-time feedback from sensors and actuators is crucial for adapting robot movements in dynamic environments, enabling robots to respond to unexpected changes and disturbances effectively.
55. How does the consideration of friction, contact forces, and compliance impact robot dynamics and control in applications like grasping and manipulation?
Answer: In tasks involving grasping and manipulation, friction, contact forces, and compliance play a significant role? Accurate modeling and control of these factors are essential for achieving successful and stable object manipulation.

56. What advancements in materials science have contributed to the development of lightweight and durable robot components, affecting kinematics and dynamics?

Answer: Innovations in materials science have led to the creation of lightweight, durable, and flexible materials that can be incorporated into robot components, influencing both kinematics and dynamics.

57. How does multi-robot coordination and collaboration rely on a deep understanding of kinematics and dynamics to achieve collective objectives?

Answer: Multi-robot coordination requires a comprehensive understanding of kinematics and dynamics to ensure that robots can work together effectively, avoiding collisions and optimizing task distribution.

58. How can robot kinematics and dynamics be used in the field of rehabilitation robotics to assist individuals with motor impairments?

Answer: Robot kinematics and dynamics are integral in designing robotic rehabilitation devices that provide tailored assistance to individuals with motor impairments, aiding in their recovery and mobility.

59. What are the key considerations in the development of bipedal robots with humanoid characteristics, and how do kinematics and dynamics play a role in their stability and agility?

Answer: Bipedal robots must replicate human-like stability and agility. Kinematics and dynamics are central in designing control systems that maintain balance and enable natural bipedal locomotion.

60. How do robot kinematics and dynamics impact the design of robotic exoskeletons for applications like medical rehabilitation and enhanced strength augmentation?

Answer: Robot kinematics and dynamics are essential for designing robotic exoskeletons that adapt to the wearer's movements while providing support and assistance. This ensures that the exoskeleton is both effective and safe.

61. In what ways do robot kinematics and dynamics contribute to the development of autonomous underwater vehicles (AUVs) for marine exploration and research?

Answer: Robot kinematics and dynamics are essential for AUVs to navigate and study underwater environments effectively. Understanding these principles allows AUVs to adapt to changing conditions and execute precise maneuvers.

62. How does the concept of energy optimization play a role in robot dynamics for applications requiring prolonged operation, such as long-duration space missions or autonomous exploration?

Answer: Energy optimization in robot dynamics is critical for extending the operational lifespan of robots in applications where power sources are limited. It involves efficient control strategies and motion planning to conserve energy.

63. What impact does robot kinematics and dynamics have on the development of swarm robotics systems for applications like environmental monitoring and agriculture?

Answer: Robot kinematics and dynamics influence the coordination and control of swarm robots. Understanding these principles helps optimize swarm behavior and collective tasks in various domains.

64. How can robot kinematics and dynamics be applied to the design of robotic hands and grippers for versatile and dexterous manipulation of objects?

Answer: Robot kinematics and dynamics are vital in designing robotic hands and grippers that can adapt to object shapes and sizes while ensuring a firm and precise grasp.

65. How does haptic feedback technology enhance human-robot interaction by conveying sensory information related to robot kinematics and dynamics?

Answer: Haptic feedback technology provides users with tactile information about robot movements and interactions. This sensory feedback enhances the intuitiveness and safety of human-robot interactions.

66. What role does model predictive control (MPC) play in optimizing robot kinematics and dynamics for tasks requiring real-time decision-making?

Answer: MPC is a control strategy that utilizes dynamic models to make real-time decisions, optimizing robot kinematics and dynamics for tasks like obstacle avoidance and trajectory tracking.

67. How does robot kinematics and dynamics contribute to the development of wearable robotic devices, such as exposits, for enhanced human performance in various fields?

Answer: Robot kinematics and dynamics are crucial in designing wearable robotic devices like exposits, ensuring that they can assist wearers in tasks requiring strength augmentation while maintaining comfort and natural movements.

68. What are the key challenges in achieving smooth and coordinated motion in multi-limbed robots, and how are these challenges addressed in their kinematics and dynamics?

Answer: Multi-limbed robots face challenges in achieving coordinated motion across their limbs. Kinematics and dynamics modeling and control strategies are used to synchronize limb movements and maintain stability.

69. How can robot kinematics and dynamics be used in the development of robots for precision agriculture, such as autonomous harvesters and crop monitoring systems?

Answer: Robot kinematics and dynamics are employed in designing agricultural robots that can navigate fields, perform precise tasks, and optimize crop management, contributing to increased agricultural efficiency.

70. What role do soft actuators and compliant mechanisms play in enhancing the flexibility and adaptability of robots in dynamic environments?

Answer: Soft actuators and compliant mechanisms provide robots with flexibility and adaptability in dynamic environments. They enable robots to interact with objects and terrain more naturally while reducing the risk of damage.

71. How can robot kinematics and dynamics contribute to the design of robotic systems for inspection and maintenance tasks in extreme environments, such as nuclear reactors and deep-sea exploration?

Answer: Robot kinematics and dynamics are essential in developing robotic systems capable of safely operating in extreme environments. They ensure precise control and stability during inspection and maintenance tasks in challenging conditions.

72. What are the implications of robot kinematics and dynamics on the development of autonomous flying robots for applications like surveillance and aerial mapping?

Answer: Robot kinematics and dynamics are critical for autonomous flying robots, ensuring stable flight, accurate navigation, and the ability to carry out aerial surveillance and mapping missions effectively.

73. How can the study of robot dynamics be applied to the development of prosthetic limbs and assistive devices that enhance the mobility and independence of individuals with limb loss?

Answer: Understanding robot dynamics plays a pivotal role in designing prosthetic limbs and assistive devices that replicate natural limb movements and provide users with enhanced mobility and independence.

74. What challenges arise in controlling the dynamics of snake-like and serpentine robots, and how are they addressed to achieve agile and flexible locomotion?

Answer: Snake-like and serpentine robots face challenges in controlling their complex dynamics. Research focuses on developing control algorithms that enable agile and flexible locomotion in these unique robotic systems.

75. How do robot kinematics and dynamics influence the development of robotic vehicles for underwater archaeology and the exploration of submerged historical sites?

Answer: Robot kinematics and dynamics are essential for underwater archaeology robots, allowing them to navigate underwater environments, document historical sites, and handle artifacts delicately.

76. What role does augmented reality (AR) play in enhancing the training and operation of robots by providing real-time visual feedback related to their kinematics and dynamics?

Answer: AR technology provides operators with real-time visual feedback about robot kinematics and dynamics, enhancing training, and assisting in robot operation by visualizing critical information and potential issues.

77. How can robot kinematics and dynamics be integrated into educational robotics programs to teach students about robotics principles and hands-on applications?

Answer: Educational robotics programs use robot kinematics and dynamics as fundamental concepts, offering students the opportunity to learn about these principles through hands-on activities and real-world applications.

78. In what ways do robot kinematics and dynamics impact the development of robotic solutions for agriculture, such as autonomous weeding robots and crop monitoring systems?

Answer: Robot kinematics and dynamics are crucial in designing agricultural robots that can navigate fields, perform precise tasks, and optimize crop management, contributing to increased agricultural efficiency.

79. How does robot kinematics and dynamics influence the design of robotic vehicles for planetary exploration and space missions?

Answer: Robot kinematics and dynamics are critical for designing robotic vehicles that can navigate the unique challenges of celestial bodies, such as low gravity and uneven surfaces. These principles ensure the reliability and efficiency of space exploration missions.

80. What role does machine learning and artificial intelligence play in enhancing robot kinematics and dynamics modeling and control, especially in complex and unstructured environments?

Answer: Machine learning and AI techniques are used to improve the accuracy and adaptability of robot kinematics and dynamics models. They enable robots to learn from data, adapt to changing environments, and optimize their movements in complex and unstructured settings.

81.How do robot kinematics and dynamics influence the design of robotic limbs for prosthetic and orthotic applications, enhancing mobility and functionality for individuals with limb loss?

Answer: Robot kinematics and dynamics are pivotal in the design of robotic limbs for prosthetic and orthotic applications. They enable these devices to replicate natural limb movements, providing enhanced mobility and functionality for individuals with limb loss. Precise control and adaptable movements are essential to help users perform daily tasks with ease.

82. What role does sensor technology, such as LiDAR and radar, play in enhancing robot kinematics and dynamics modeling for autonomous vehicles, ensuring safe and efficient navigation on roads and in urban environments?

Answer: Sensor technology, including LiDAR and radar, is instrumental in improving robot kinematics and dynamics modeling for autonomous vehicles. These sensors provide real-time data about the environment, helping autonomous vehicles navigate safely and efficiently on roads and in urban environments. Accurate perception of obstacles, pedestrians, and traffic conditions allows for dynamic adjustments in vehicle motion planning.

83. How can robot kinematics and dynamics contribute to the development of swarm robots for agricultural tasks, optimizing crop management and yield through coordinated actions in large-scale farming operations?

Answer: Robot kinematics and dynamics are crucial in designing swarm robots for agricultural tasks. They enable precise coordination and synchronized actions among robots, optimizing crop management and yield in large-scale farming operations. These robots can work together efficiently, performing tasks such as planting, harvesting, and pest control, ultimately enhancing agricultural productivity.

84. What are the implications of robot kinematics and dynamics in the design and control of robotic arms used in hazardous environments, such as nuclear facilities and chemical plants, for inspection and maintenance purposes?

Answer: Robot kinematics and dynamics have significant implications in the design and control of robotic arms for hazardous environments. These principles ensure precise and stable control, allowing robots to perform inspection and maintenance tasks in nuclear facilities and chemical plants safely. Accurate manipulation and tool operation are essential to prevent accidents and protect human operators.

85. How does the concept of redundancy resolution in robot kinematics impact the development of redundant manipulators used in research, manufacturing, and medical applications, ensuring optimal performance while avoiding joint limits and singularities?

Answer: Redundancy resolution is crucial in developing redundant manipulators used in various applications. It allows these manipulators to optimize their movements, ensuring optimal performance while avoiding joint limits and singularities. In research, manufacturing, and medical applications, this capability enhances precision and flexibility, enabling robots to perform tasks efficiently and adapt to changing conditions.

86. What role does energy efficiency play in robot dynamics modeling and control, especially in applications where energy conservation is essential, such as renewable energy production and autonomous vehicles?
Answer: Energy efficiency is a critical consideration in robot dynamics modeling and control, particularly in applications where energy conservation is essential. In renewable energy production and autonomous vehicles, efficient control strategies and motion planning are employed to minimize energy consumption. This helps extend operational duration and reduce the environmental impact of these systems.

ONLINE RESOURCES

Practice Paper Link: https://forms.gle/q3iUtfNDvUnd2Tgo8

QR Code:

Video Content Link:
https://www.youtube.com/playlist?list=PLzlh_31VWEFqItZJm05ydCJaXL8Z8mrtJ

QR Code:

Chapter 5: Robot Sensors

1. What are the primary types of sensors commonly used in robotics?

Answer: The primary types of sensors in robotics include proximity sensors, vision sensors, tactile sensors, and inertial sensors.

2. How do proximity sensors work, and what are their applications in robotics?

Answer: Proximity sensors detect the presence or absence of objects by emitting a signal and measuring its reflection. They are used for object detection, collision avoidance, and robot navigation.

3. Explain the role of vision sensors in robotic perception.

Answer: Vision sensors capture visual data to help robots understand their surroundings. They are vital for tasks like object recognition, localization, and mapping.

4. What are tactile sensors, and how do they enable robots to interact with objects?

Answer: Tactile sensors measure contact forces and pressure, allowing robots to interact with objects delicately. They are used in applications such as grasping, manipulation, and human-robot collaboration.

5. How do inertial sensors contribute to robot motion control and stability?

Answer: Inertial sensors, like accelerometers and gyroscopes, measure changes in acceleration and orientation. They play a key role in robot motion control, stability, and navigation.

6. What is the purpose of force/torque sensors in robotics, and where are they commonly applied?

Answer: Force/torque sensors measure the forces and torques applied to a robot's end-effector. They are used in applications like assembly tasks, quality control, and haptic feedback.

7. How do ultrasonic sensors function, and what are their advantages in robotics?

Answer: Ultrasonic sensors use sound waves to measure distances. They are advantageous for detecting objects, obstacles, and proximity in robotics, especially in challenging environments.

8. Explain the concept of LIDAR (Light Detection and Ranging) sensors and their significance in robot perception.

Answer: LIDAR sensors use lasers to measure distances and create 3D maps of the environment. They are crucial for autonomous navigation, mapping, and obstacle avoidance.

9. What is the role of thermal sensors in robotics, and where are they commonly employed?

Answer: Thermal sensors detect heat signatures and temperature variations. They find applications in areas such as search and rescue, surveillance, and detecting anomalies in machinery.

10. How can magnetic sensors be used in robotics, and what are their advantages?

Answer: Magnetic sensors detect changes in magnetic fields. They are employed in tasks like precise positioning, navigation, and tracking using magnets or magnetic landmarks.

11. What are the benefits of using optical encoders in robotic joint control and positioning?

Answer: Optical encoders provide precise feedback on joint angles and positions, aiding in robot joint control, kinematics, and accurate positioning.

12. How do capacitive sensors work, and where are they utilized in robotics?

Answer: Capacitive sensors measure changes in capacitance caused by proximity or touch. They are used for object detection, material identification, and touch-sensitive applications.

13. Explain the concept of Hall Effect sensors and their applications in robotics.

Answer: Hall Effect sensors detect magnetic field variations and find applications in motor control, position sensing, and detecting the presence of magnets in robotic systems.

14. What is the role of pressure sensors in robotics, and where can they be applied?

Answer: Pressure sensors measure fluid or gas pressure and are used in applications like robotic grippers, industrial automation, and monitoring fluid systems.

15. How do GPS (Global Positioning System) sensors contribute to outdoor robot navigation?

Answer: GPS sensors determine a robot's precise location and are essential for outdoor navigation, especially in autonomous vehicles and drones.

16. What is the significance of environmental sensors in robotics, and where are they commonly used?

Answer: Environmental sensors measure parameters like temperature, humidity, and air quality. They are employed in applications such as environmental monitoring, agriculture, and indoor air quality control.

17. How can chemical sensors be applied in robotic systems, and what types of substances can they detect?

Answer: Chemical sensors identify specific chemicals or gases and are used in applications like gas leak detection, environmental analysis, and industrial safety.

18. Explain the concept of RGB-D sensors and their role in robot perception.

Answer: RGB-D sensors combine color (RGB) and depth (D) information to create detailed 3D maps of the environment. They are valuable for object recognition, scene understanding, and obstacle avoidance in robotics.

19. What are the advantages of using radar sensors in autonomous vehicles and robotics?

Answer: Radar sensors provide long-range object detection and are valuable for applications like autonomous driving, collision avoidance, and surveillance in adverse weather conditions.

20. How do time-of-flight (Tofu) sensors work, and what are their applications in robotics?

Answer: Time-of-flight sensors measure the time it takes for light to travel to an object and back, enabling accurate distance measurements. They are used in robotics for obstacle detection, gesture recognition, and depth sensing.

21. What are the challenges associated with sensor fusion in robotics, and how does it enhance perception and decision-making?

Answer: Sensor fusion involves integrating data from multiple sensors to improve accuracy and reliability. Challenges include data alignment and synchronization. Sensor fusion enhances perception and decision-making in robotics by providing a comprehensive and robust understanding of the environment.

22. How can robots use environmental sensors to adapt to changing conditions and optimize their performance?

Answer: Robots use environmental sensors to gather real-time data about their surroundings. This information helps them adapt to changing conditions, such as lighting, temperature, or humidity, optimizing their performance in various tasks and environments.

23. Explain the concept of active vs. passive sensors in robotics, and provide examples of each.

Answer: Active sensors emit a signal (e.g., ultrasound, laser) and measure its reflection, while passive sensors detect existing signals (e.g., light, sound). Active sensor examples include LIDAR, ultrasonic sensors, and radar, while passive sensors include cameras and microphones.

24. What is the role of sensor calibration in robotics, and why is it essential for accurate robot perception?

Answer: Sensor calibration involves adjusting sensor measurements to eliminate inaccuracies. It is essential for accurate robot perception because it ensures that sensor data aligns correctly with the physical environment, enabling precise decision-making and control.

25. How can robots utilize sensor data fusion to improve their understanding of complex environments and situations?

Answer: Sensor data fusion combines information from multiple sensors to create a more comprehensive and accurate representation of the environment. This improves a robot's understanding of complex situations, enhancing its ability to make informed decisions and navigate effectively.

26. What are the privacy and ethical considerations associated with the use of sensors in robotics, particularly in applications like surveillance and human-robot interaction?

Answer: Privacy and ethical considerations include data security, consent, and the responsible use of sensor data. These issues must be addressed when deploying sensors in applications involving surveillance and human-robot interaction to ensure privacy and respect ethical guidelines.

27. How do environmental conditions, such as extreme temperatures or adverse weather, affect the performance and reliability of sensors in robotics?

Answer: Extreme temperatures and adverse weather conditions can impact sensor performance and reliability. Sensors may become less accurate or fail in extreme environments, necessitating specialized sensor designs or protective measures for robotic systems.

28. What are some emerging sensor technologies that have the potential to revolutionize robotics in the future?

Answer: Emerging sensor technologies include quantum sensors, bio-inspired sensors, and neuromorphic sensors, which offer advanced capabilities such as ultra-precise measurements, biomimetic sensing, and brain-inspired perception, potentially revolutionizing the field of robotics.

29. How can robots benefit from advancements in wireless communication and sensor networks to share data and collaborate in real-time?

Answer: Robots can benefit from wireless communication and sensor networks by sharing data and collaborating in real-time. This enables coordinated actions among robots, facilitating tasks like search and rescue, exploration, and surveillance in distributed environments.

30. What are the key considerations for selecting the appropriate sensors for a specific robotics application, and how can sensor integration be optimized for performance?

Answer: Key considerations for sensor selection include the type of data needed, environmental conditions, and budget constraints. Optimizing sensor integration involves careful calibration, synchronization, and data processing to ensure reliable and high-performance sensor functionality in the chosen application.

31. How do 3D depth sensors like Microsoft Kinect contribute to human-robot interaction and gesture recognition?

Answer: 3D depth sensors like Microsoft Kinect enable robots to perceive and interact with humans in a more intuitive way. They accurately capture human movements and gestures, making human-robot interaction natural and responsive. These sensors find applications in gaming, healthcare, and robotics, where they enhance gesture-based control and communication.

32. What are the advantages of using gas sensors in robotics for applications like environmental monitoring and air quality assessment?

Answer: Gas sensors are advantageous in robotics for environmental monitoring and air quality assessment. They can detect various gases and pollutants, providing crucial data for assessing environmental conditions and ensuring safe and healthy surroundings in applications ranging from smart cities to industrial safety.

33. How can lidar sensors be applied in autonomous navigation for robots, drones, and self-driving vehicles, enabling safe and precise movement in complex environments?

Answer: Lidar sensors are essential for autonomous navigation in robots, drones, and self-driving vehicles. They create detailed 3D maps of the surroundings by emitting laser pulses and measuring their reflections. These maps enable robots to navigate safely in complex environments, avoid obstacles, and make informed decisions, making them crucial for autonomy and safety.

34. Explain the role of Hall Effect sensors in motor control and the position feedback of robotic joints.

Answer: Hall Effect sensors play a vital role in motor control and position feedback of robotic joints. By measuring changes in magnetic fields, they provide accurate data about joint positions, allowing precise control of robotic movements and ensuring accurate positioning in applications like industrial automation and robotics.

35. What is the significance of thermal imaging sensors in robotics, particularly in search and rescue operations and detecting hidden objects?

Answer: Thermal imaging sensors are significant in robotics for search and rescue operations and detecting hidden objects. They capture heat signatures, making it possible to locate individuals in disaster scenarios, find concealed objects, and enhance situational awareness in challenging environments.

36. How can humidity sensors be utilized in agricultural robots to monitor soil moisture levels and optimize irrigation practices?

Answer: Humidity sensors are used in agricultural robots to monitor soil moisture levels and optimize irrigation practices. They provide real-time data about soil humidity, enabling precise control of watering systems. This ensures efficient water usage and promotes healthy crop growth, making them valuable in precision agriculture.

37. Explain the role of bio-inspired sensors in robotics and how they mimic natural systems to enhance robot perception and adaptability.

Answer: Bio-inspired sensors mimic natural systems in robotics to enhance perception and adaptability. These sensors draw inspiration from biological organisms to replicate their sensory capabilities, enabling robots to perceive and interact with their environments more effectively. Examples include sensors inspired by insects' compound eyes and tactile sensors resembling human skin.

38. What are the applications of acoustic sensors in robotics, including underwater robots, autonomous vehicles, and surveillance systems?

Answer: Acoustic sensors have diverse applications in robotics, including underwater robots for sonar-based navigation, autonomous vehicles for collision avoidance and object detection, and surveillance systems for audio monitoring and intrusion detection.

39. How can robots benefit from using MEMS (Micro-Electro-Mechanical Systems) sensors for compact and high-performance sensing solutions?

Answer: Robots benefit from MEMS sensors due to their compact size and high performance. MEMS sensors include accelerometers, gyroscopes, and pressure sensors, enabling robots to have precise motion control, navigation, and environmental sensing in a compact form factor.

40. What are the ethical considerations regarding the use of sensor data in robotics, and how can responsible data handling be ensured in various robotic applications?

Answer: Ethical considerations in sensor data use involve privacy, consent, and data security. Responsible data handling in robotics requires clear data policies, secure storage, and compliance with regulations. It also necessitates transparency in how sensor data is collected, processed, and shared to ensure ethical and responsible use in applications like surveillance, healthcare, and human-robot interaction.

41. How do magnetic field sensors contribute to robotics in applications such as robot localization, mapping, and magnetically-guided navigation?

Answer: Magnetic field sensors contribute to robotics by detecting changes in magnetic fields. They play a crucial role in robot localization, mapping, and magnetically-guided navigation, especially in environments where traditional positioning methods may be less effective. These sensors enable robots to determine their positions accurately and navigate along predefined magnetic paths.

42. What are the advantages of using biosensors in robotics for applications like healthcare, diagnostics, and environmental monitoring?

Answer: Biosensors offer advantages in robotics for healthcare, diagnostics, and environmental monitoring. They can detect specific biological molecules, pathogens, or toxins, providing valuable information for medical diagnosis, disease monitoring, and environmental analysis. Biosensors enhance the capabilities of robotic systems in fields where biological sensing is essential.

43. How do ultraviolet (UV) sensors enhance robot safety and performance in applications like UV disinfection and solar energy harvesting?

Answer: UV sensors enhance robot safety and performance in applications like UV disinfection and solar energy harvesting. They detect UV radiation, ensuring that robots operate safely in UV-intensive environments and optimize energy harvesting from solar sources, contributing to improved efficiency and effectiveness.

44. Explain the concept of piezoelectric sensors and their applications in robotics for detecting vibrations, pressure, and mechanical stress.

Answer: Piezoelectric sensors generate electrical signals when subjected to mechanical stress, vibrations, or pressure. They are used in robotics to detect and measure these physical phenomena, finding applications in structural health monitoring, vibration analysis, and tactile feedback in human-robot interaction.

45. What role do color sensors play in robotics for tasks like object recognition, sorting, and quality control in manufacturing processes?

Answer: Color sensors are essential in robotics for tasks like object recognition, sorting, and quality control in manufacturing processes. They capture and analyze color information, enabling robots to identify objects, categorize them, and ensure product quality by inspecting colors and patterns.

46. How can pressure-sensitive sensors be employed in robotics to enhance safety and collision detection in human-robot collaboration scenarios?

Answer: Pressure-sensitive sensors are employed in robotics to enhance safety in human-robot collaboration scenarios. They can detect contact and pressure changes, allowing robots to stop or adjust their movements when they encounter obstacles or interact with humans, reducing the risk of collisions and injuries.

47. What are the key considerations when using radiation sensors in robotics for applications like nuclear plant inspection and radiation monitoring in hazardous environments?

Answer: When using radiation sensors in robotics for hazardous applications, key considerations include calibration, shielding, and radiation-resistant materials. Proper maintenance and safety protocols must be followed to ensure accurate measurements and protect robots and operators from radiation exposure.

48. How do robots utilize GPS (Global Positioning System) sensors in outdoor navigation, mapping, and geolocation-based tasks?

Answer: Robots utilize GPS sensors for outdoor navigation, mapping, and geolocation-based tasks. GPS provides accurate positioning information, allowing robots to determine their coordinates, plan routes, and perform tasks like mapping, geofencing, and location-based data collection.

49. What are the advantages of using humidity sensors in climate-controlled environments, such as greenhouses and controlled-atmosphere storage facilities?

Answer: Humidity sensors offer advantages in climate-controlled environments like greenhouses and controlled-atmosphere storage facilities. They help maintain optimal humidity levels for plant growth, crop preservation, and quality control, ensuring ideal conditions in agricultural and storage applications.

50. How can chemical sensors be applied in the detection of hazardous materials and pollutants in industrial settings and environmental monitoring?

Answer: Chemical sensors are valuable in industrial settings and environmental monitoring for detecting hazardous materials and pollutants. They identify specific chemicals or gases, enabling early detection of toxic substances, ensuring workplace safety, and contributing to environmental protection and air quality assessment.

51. What is the role of laser-based rangefinders in robotics and how do they enable accurate distance measurements in various applications?

Answer: Laser-based rangefinders are pivotal in robotics for accurate distance measurements. They emit laser beams and measure the time it takes for the light to bounce back, allowing robots to determine distances precisely. These sensors find applications in tasks like object detection, mapping, and obstacle avoidance, enabling robots to navigate and interact with their environment with high precision.

52. How do gas chromatography sensors contribute to the detection and analysis of chemical compounds in robotic systems, particularly in fields like environmental monitoring and forensic science?

Answer: Gas chromatography sensors are instrumental in detecting and analyzing chemical compounds in robotic systems. They separate and identify complex mixtures of gases, making them valuable in environmental monitoring for pollutant analysis and in forensic science for identifying trace substances at crime scenes.

53. Explain the significance of proximity capacitive sensors in robotic touch-sensitive interfaces and object detection applications.

Answer: Proximity capacitive sensors are significant in robotics for touch-sensitive interfaces and object detection. They detect changes in capacitance caused by nearby objects or touch, making them ideal for applications like touchscreens, proximity detection, and human-robot interaction, where precise and non-contact sensing is required.

54. How do radar sensors enable robotics to detect and track moving objects, making them suitable for applications like autonomous vehicles and surveillance systems?

Answer: Radar sensors enable robotics to detect and track moving objects by emitting radio waves and measuring their reflections. They are suitable for applications like autonomous vehicles and surveillance systems, where reliable detection of moving targets, accurate velocity measurements, and long-range sensing are crucial.

55. What are the advantages of using photoelectric sensors in robotics for tasks like object counting, position sensing, and industrial automation?

Answer: Photoelectric sensors offer advantages in robotics for various tasks, including object counting, position sensing, and industrial automation. They use light beams to detect objects, providing reliable detection and precise position information, making them versatile in manufacturing and robotics applications.

56. How can robots employ biosensors for healthcare applications, including monitoring vital signs, diagnosing diseases, and drug delivery?

Answer: Robots can employ biosensors in healthcare applications for monitoring vital signs, diagnosing diseases, and drug delivery. Biosensors detect biological markers, enabling robots to gather health-related data, diagnose conditions, and administer medications with precision, contributing to improved patient care and medical automation.

57. Explain the concept of radiation-tolerant sensors and their significance in space exploration, nuclear applications, and radiation therapy in robotics.

Answer: Radiation-tolerant sensors can withstand ionizing radiation, making them crucial in space exploration, nuclear applications, and radiation therapy in robotics. They provide reliable data in high-radiation environments, ensuring the safe operation of robots and enhancing their capabilities in these critical fields.

58. What role do force-sensitive resistors (FSRs) play in robotic grippers and prosthetic devices, enhancing grip control and tactile feedback?

Answer: Force-sensitive resistors (FSRs) play a vital role in robotic grippers and prosthetic devices by providing grip control and tactile feedback. They respond to applied pressure, allowing robots and prosthetics to grasp objects with the right force and providing users with a sense of touch and object manipulation.

59. How can robots utilize humidity and temperature sensors in climate control and precision agriculture, optimizing conditions for plant growth and crop management?

Answer: Robots can utilize humidity and temperature sensors in climate control and precision agriculture. These sensors monitor environmental conditions, allowing precise control of temperature and humidity in greenhouses and agricultural settings. This optimization promotes healthy plant growth, efficient crop management, and resource conservation.

60. What is the significance of photodetectors in robotics for tasks like light sensing, object detection, and optical communication?

Answer: Photodetectors are significant in robotics for light sensing, object detection, and optical communication. They convert light into electrical signals, enabling robots to perceive light levels, detect objects based on reflections, and establish optical communication links, expanding their capabilities in various applications.

61. How can robots employ load cells for measuring force and weight in industrial automation, material testing, and quality control processes?

Answer: Robots can employ load cells for measuring force and weight in industrial automation, material testing, and quality control processes. Load cells provide accurate force measurements, ensuring proper handling of materials, product quality assessment, and precise control in manufacturing and robotics applications.

62. What are the applications of color temperature sensors in robotics, particularly in tasks related to color consistency, lighting control, and photography?

Answer: Color temperature sensors find applications in robotics for tasks related to color consistency, lighting control, and photography. They measure the color temperature of light sources, ensuring consistent lighting conditions, accurate color rendering, and optimal image capture in photography and color-sensitive applications.

63. How do oxygen sensors contribute to robotics for monitoring oxygen levels in controlled environments, medical devices, and underwater exploration?

Answer: Oxygen sensors contribute to robotics by monitoring oxygen levels in controlled environments, medical devices, and underwater exploration. They detect oxygen concentrations, ensuring safety, health monitoring, and oxygen management in various applications, from controlled atmospheres to healthcare and marine robotics.

64. Explain the role of vibration sensors in robotics for structural health monitoring, condition-based maintenance, and machinery fault detection.

Answer: Vibration sensors play a critical role in robotics for structural health monitoring, condition-based maintenance, and machinery fault detection. They detect vibrations in equipment and structures, enabling early fault detection, predictive maintenance, and ensuring the safety and reliability of robotic systems.

65. What are the advantages of using gyroscopic sensors in robotic navigation, orientation tracking, and motion stabilization, especially in drones and UAVs?

Answer: Gyroscopic sensors offer advantages in robotic navigation, orientation tracking, and motion stabilization, particularly in drones and UAVs. They measure angular velocity and provide precise orientation data, ensuring stable flight, accurate navigation, and responsive motion control in aerial robotics.

66. How can robots utilize current sensors for monitoring electrical currents in industrial automation, energy management, and motor control applications?

Answer: Robots can utilize current sensors for monitoring electrical currents in industrial automation, energy management, and motor control applications. These sensors provide real-time data on current flow, enabling efficient energy usage, motor protection, and process control in manufacturing and robotics.

67. What are the applications of magneto resistive sensors in robotics for tasks like magnetic field mapping, navigation, and position sensing in autonomous systems?

Answer: Magneto resistive sensors have applications in robotics for magnetic field mapping, navigation, and position sensing in autonomous systems. They detect changes in magnetic fields, allowing robots to create maps, determine positions, and navigate using magnetic landmarks in challenging environments.

68. How do light curtains contribute to robot safety in industrial settings, providing reliable intrusion detection and collision prevention?

Answer: Light curtains enhance robot safety in industrial settings by providing reliable intrusion detection and collision prevention. They create protective barriers of light beams that, when interrupted, trigger immediate safety measures, ensuring the safety of personnel and preventing accidents in robotic work zones.

69. Explain the significance of humidity-compensated temperature sensors in robotics for accurate temperature measurements in humid environments.

Answer: Humidity-compensated temperature sensors are significant in robotics for accurate temperature measurements in humid environments. They adjust temperature readings based on humidity levels, ensuring precise data even in conditions where humidity can affect temperature measurements, such as HVAC control and weather monitoring.

70. How can robots benefit from using time-of-flight (Tofu) distance sensors in gesture recognition, obstacle detection, and virtual reality applications?

Answer: Robots benefit from using time-of-flight (Tofu) distance sensors in gesture recognition, obstacle detection, and virtual reality applications. Tofu sensors measure distances with high accuracy and low latency, making them ideal for real-time gesture tracking, obstacle avoidance, and immersive virtual reality experiences.

71. What is the role of radio frequency identification (RFID) sensors in robotics for inventory management, asset tracking, and access control?

Answer: RFID sensors play a vital role in robotics for inventory management, asset tracking, and access control. They use radio waves to identify and track objects or assets, improving efficiency in logistics, inventory control, and secure access systems.

72. How can robots employ carbon dioxide (CO2) sensors for environmental monitoring, indoor air quality assessment, and climate control?

Answer: Robots can employ carbon dioxide (CO2) sensors for environmental monitoring, indoor air quality assessment, and climate control. These sensors measure CO2 levels, allowing precise control of ventilation and climate systems, promoting healthy indoor environments and reducing energy consumption.

73. Explain the concept of piezo resistive sensors and their applications in robotics for measuring pressure, force, and strain in various industrial and medical devices.

Answer: Piezo resistive sensors measure pressure, force, and strain by detecting changes in electrical resistance. They find applications in robotics for precise pressure sensing in medical devices like blood pressure monitors, force measurements in robotic grippers, and strain monitoring in industrial equipment, ensuring accurate data and control.

74. What are the advantages of using thermal cameras and infrared sensors in robotics for temperature measurement, thermal imaging, and security applications?

Answer: Thermal cameras and infrared sensors offer advantages in robotics for temperature measurement, thermal imaging, and security applications. They capture heat signatures, enabling non-contact temperature measurements, night vision capabilities, and enhanced security surveillance in robotics and security systems.

75. How do robots employ position encoders for accurate feedback in applications like robotic arms, CNC machines, and 3D printers?

Answer: Robots employ position encoders for accurate feedback in applications like robotic arms, CNC machines, and 3D printers. These encoders provide precise position data, ensuring accurate control, motion planning, and repeatability in various robotic and automated systems.

76. What are the applications of tilt sensors in robotics for detecting changes in orientation, leveling, and platform stabilization in mobile robots and drones?

Answer: Tilt sensors have applications in robotics for detecting changes in orientation, leveling, and platform stabilization in mobile robots and drones. They measure inclination angles, allowing robots to maintain stability, adjust orientation, and navigate challenging terrains with precision.

77. How can robots utilize pH sensors in laboratory automation, water quality monitoring, and chemical analysis applications?

Answer: Robots can utilize pH sensors in laboratory automation, water quality monitoring, and chemical analysis applications. PH sensors measure acidity or alkalinity levels, ensuring accurate and automated pH control, sample testing, and analysis in various scientific and industrial processes.

78. Explain the role of fiber optic sensors in robotics for tasks like strain measurement, structural health monitoring, and remote sensing in harsh environments.

Answer: Fiber optic sensors play a vital role in robotics for tasks like strain measurement, structural health monitoring, and remote sensing in harsh environments. They use optical fibers to detect changes in strain, temperature, or environmental conditions, ensuring reliable data collection and monitoring in challenging and remote locations.

79. What is the significance of air quality sensors in robotics for assessing indoor air quality, pollutant detection, and environmental health monitoring?

Answer: Air quality sensors are significant in robotics for assessing indoor air quality, pollutant detection, and environmental health monitoring. They measure air pollutants, providing valuable data for indoor air quality management, pollution control, and environmental health assessment.

80. How do robots benefit from using ultraviolet (UV) sensors for UV radiation measurement, sterilization, and monitoring in UV-sensitive applications?

Answer: Robots benefit from using ultraviolet (UV) sensors for UV radiation measurement, sterilization, and monitoring in UV-sensitive applications. UV sensors measure UV intensity, enabling precise UV exposure control, sterilization processes, and UV-sensitive equipment monitoring, ensuring safety and quality in various applications.

81. What are the applications of proximity inductive sensors in robotics, including object detection, automation, and machinery control?

Answer: Proximity inductive sensors find applications in robotics for object detection, automation, and machinery control. They use electromagnetic fields to detect the presence of metal objects, making them suitable for tasks like detecting metallic parts, controlling conveyor systems, and automating manufacturing processes.

82. How can robots employ water quality sensors for monitoring water parameters in aquatic environments, water treatment, and environmental research?

Answer: Robots can employ water quality sensors for monitoring water parameters in aquatic environments, water treatment, and environmental research. These sensors measure parameters such as pH, dissolved oxygen, and turbidity, ensuring accurate water quality assessment, environmental monitoring, and water treatment process control.

83. Explain the significance of radio sensors (RF sensors) in robotics for remote sensing, data transmission, and tracking applications.

Answer: Radio sensors (RF sensors) are significant in robotics for remote sensing, data transmission, and tracking applications. They use radio frequency signals to gather data remotely, establish communication links, and track objects or devices, enhancing the capabilities of robots in remote sensing and wireless communication.

84. How do robots utilize position-sensitive detectors (PSDs) for precise position measurement in applications like optical alignment, tracking, and laser beam control?

Answer: Robots utilize position-sensitive detectors (PSDs) for precise position measurement in applications like optical alignment, tracking, and laser beam control. PSDs detect the position of incident light, enabling accurate alignment, tracking of moving objects, and laser beam steering in various optical and robotics applications.

85. What are the advantages of using humidity and temperature combined sensors (hygrometers) in robotics for climate control, environmental monitoring, and HVAC systems?

Answer: Humidity and temperature combined sensors (hygrometers) offer advantages in robotics for climate control, environmental monitoring, and HVAC systems. They measure both humidity and temperature, providing comprehensive environmental data for precise climate control, indoor air quality assessment, and energy-efficient HVAC systems.

86. How can robots employ gas flow sensors for monitoring gas flow rates, leak detection, and gas handling in industrial and laboratory settings?

Answer: Robots can employ gas flow sensors for monitoring gas flow rates, leak detection, and gas handling in industrial and laboratory settings. These sensors measure gas flow accurately, ensuring efficient gas control, leak detection, and process monitoring in various applications.

87. Explain the role of strain gauges in robotics for measuring mechanical deformation, stress, and load in structures and robotic components.

Answer: Strain gauges play a critical role in robotics for measuring mechanical deformation, stress, and load in structures and robotic components. They detect changes in electrical resistance caused by strain, enabling precise strain measurement, structural health monitoring, and load analysis in robotics and engineering.

88. What are the applications of tilt-compensated compass sensors in robotics for accurate orientation determination, navigation, and heading control?

Answer: Tilt-compensated compass sensors have applications in robotics for accurate orientation determination, navigation, and heading control. They compensate for tilt or movement, ensuring reliable heading information, precise navigation, and orientation control in robotics and autonomous systems.

89. How can robots utilize passive infrared (PIR) motion sensors for human detection, security monitoring, and energy-saving applications in smart homes and buildings?

Answer: Robots can utilize passive infrared (PIR) motion sensors for human detection, security monitoring, and energy-saving applications in smart homes and buildings. PIR sensors detect changes in infrared radiation caused by motion, enabling human presence detection, security alerts, and automated lighting control, contributing to home automation and energy efficiency.

90. Explain the significance of light-to-frequency sensors (LTF) in robotics for light intensity measurement, color detection, and optical sensing.

Answer: Light-to-frequency sensors (LTF) are significant in robotics for light intensity measurement, color detection, and optical sensing. They convert light intensity into frequency signals, allowing robots to measure light levels, detect colors, and perform optical sensing tasks with precision and versatility.

91. What role do displacement sensors play in robotics for precise position feedback, dimensional measurement, and calibration in manufacturing and metrology?

Answer: Displacement sensors play a crucial role in robotics for precise position feedback, dimensional measurement, and calibration in manufacturing and metrology. They measure changes in position with high accuracy, ensuring precise control, quality assurance, and dimensional analysis in robotic and metrology applications.

92. How can robots employ ion-selective sensors for detecting specific ions in chemical analysis, environmental monitoring, and water quality assessment?

Answer: Robots can employ ion-selective sensors for detecting specific ions in chemical analysis, environmental monitoring, and water quality assessment. These sensors selectively detect ions, providing valuable data for chemical analysis, pollution control, and water quality management in various applications.

93. Explain the concept of load cell sensors and their applications in robotics for weight measurement, force feedback, and material testing.

Answer: Load cell sensors measure weight and force by detecting deformation in their structures. In robotics, they find applications in weight measurement for logistics, force feedback for haptic interfaces, and material testing for quality control and research purposes.

94. What are the advantages of using ultrasonic distance sensors in robotics for non-contact distance measurement, obstacle detection, and navigation?

Answer: Ultrasonic distance sensors offer advantages in robotics for non-contact distance measurement, obstacle detection, and navigation. They emit and receive ultrasonic waves to determine distances accurately, making them suitable for tasks like obstacle avoidance, navigation, and proximity sensing in robotics.

95. How do robots employ potentiometers for rotary position sensing and feedback in applications like robotic joints, servos, and control systems?

Answer: Robots employ potentiometers for rotary position sensing and feedback in applications like robotic joints, servos, and control systems. Potentiometers measure angular positions, providing feedback for precise control, motion planning, and positional accuracy in various robotic systems.

96. What is the significance of pressure transducers in robotics for measuring fluid pressure, hydraulic control, and process automation?

Answer: Pressure transducers are significant in robotics for measuring fluid pressure, hydraulic control, and process automation. They convert fluid pressure into electrical signals, enabling accurate pressure measurement, control, and automation in hydraulic systems and industrial processes.

97. How can robots utilize soil moisture sensors in agriculture for real-time monitoring, irrigation control, and crop optimization?

Answer: Robots can utilize soil moisture sensors in agriculture for real-time monitoring, irrigation control, and crop optimization. These sensors measure soil moisture levels, allowing precise irrigation management, water conservation, and improved crop health in precision agriculture.

98. Explain the role of laser range sensors (LRS) in robotics for 2D and 3D mapping, object detection, and localization in autonomous systems.

Answer: Laser range sensors (LRS) play a critical role in robotics for 2D and 3D mapping, object detection, and localization in autonomous systems. They emit laser beams to measure distances and create detailed maps, enabling robots to detect objects, navigate, and localize themselves accurately in complex environments.

99. What are the applications of ultraviolet (UV) detectors in robotics for UV radiation measurement, sterilization, and UV-sensitive applications like scientific experiments?

Answer: Ultraviolet (UV) detectors have applications in robotics for UV radiation measurement, sterilization, and UV-sensitive applications like scientific experiments. They measure UV intensity, ensuring precise UV exposure control, sterilization processes, and UV-sensitive experiment monitoring in robotics and scientific research.

100. How do robots employ force-torque sensors for grasping, manipulation, and fine motor skills, enhancing their dexterity and adaptability in various tasks?

Answer: Robots employ force-torque sensors for grasping, manipulation, and fine motor skills, enhancing their dexterity and adaptability. These sensors measure contact forces and torques, allowing robots to grasp objects delicately, adjust their grip, and perform tasks that require precision and sensitivity in various applications.

101. What are the applications of laser displacement sensors in robotics for measuring distance and dimensions in industrial quality control, machining, and 3D scanning?

Answer: Laser displacement sensors find applications in robotics for measuring distance and dimensions in industrial quality control, machining, and 3D scanning. They use laser beams to precisely measure distances and dimensions, ensuring quality control, accurate machining, and detailed 3D scans in manufacturing and metrology.

102. How can robots utilize methane sensors for gas detection, safety monitoring, and environmental analysis in applications like natural gas pipelines and underground mining?

Answer: Robots can utilize methane sensors for gas detection, safety monitoring, and environmental analysis in applications like natural gas pipelines and underground mining. Methane sensors detect methane gas levels, ensuring safety, leak detection, and environmental protection in potentially hazardous environments.

103. Explain the role of acoustic emission sensors in robotics for structural health monitoring, crack detection, and material characterization in aerospace and civil engineering.

Answer: Acoustic emission sensors play a crucial role in robotics for structural health monitoring, crack detection, and material characterization in aerospace and civil engineering. They detect acoustic signals generated by cracks or defects, enabling early fault detection, quality assurance, and structural integrity assessment in critical applications.

104. What are the advantages of using force-sensitive resistors (FSRs) in robotic foot sensors for terrain adaptation, gait control, and stability enhancement in legged robots?

Answer: Force-sensitive resistors (FSRs) offer advantages in robotic foot sensors for terrain adaptation, gait control, and stability enhancement in legged robots. They detect ground contact forces, allowing robots to adjust their gait, navigate uneven terrain, and maintain stability, improving locomotion in legged robotics.

105. How do robots employ oxygen sensors for monitoring oxygen levels in enclosed spaces, oxygen therapy equipment, and life support systems?

Answer: Robots employ oxygen sensors for monitoring oxygen levels in enclosed spaces, oxygen therapy equipment, and life support systems. These sensors measure oxygen concentrations, ensuring safety, oxygen therapy control, and reliable life support in medical and confined space applications.

106. What is the significance of gas chromatography-mass spectrometry (GC-MS) sensors in robotics for chemical analysis, environmental monitoring, and forensic investigations?

Answer: Gas chromatography-mass spectrometry (GC-MS) sensors are significant in robotics for chemical analysis, environmental monitoring, and forensic investigations. They separate and analyze complex mixtures of gases and compounds, providing valuable data for chemical identification, pollution control, and forensic analysis in various applications.

107. How can robots utilize radar altimeters for precise altitude measurements, terrain following, and obstacle avoidance in aerial and autonomous systems?

Answer: Robots can utilize radar altimeters for precise altitude measurements, terrain following, and obstacle avoidance in aerial and autonomous systems. Radar altimeters use radio waves to determine altitude above ground level, ensuring accurate altitude control, terrain awareness, and obstacle avoidance in robotics and aviation.

108. Explain the role of spectral sensors in robotics for spectral analysis, material identification, and remote sensing in agriculture, geology, and environmental science.

Answer: Spectral sensors play a vital role in robotics for spectral analysis, material identification, and remote sensing in agriculture, geology, and environmental science. They capture spectral data from surfaces, enabling material identification, vegetation analysis, and environmental monitoring in various scientific and agricultural applications.

109. What are the applications of colorimeters in robotics for color measurement, quality control, and color matching in industries like textiles, cosmetics, and printing?

Answer: Colorimeters have applications in robotics for color measurement, quality control, and color matching in industries like textiles, cosmetics, and printing. They measure color values accurately, ensuring consistent color quality, product inspection, and color matching in manufacturing and quality control processes.

110. How can robots employ strain gauge sensors for measuring strain and deformation in structures, load-bearing components, and stress analysis in engineering applications?

Answer: Robots can employ strain gauge sensors for measuring strain and deformation in structures, load-bearing components, and stress analysis in engineering applications. Strain gauge sensors detect changes in resistance due to strain, providing crucial data for structural integrity assessment, load monitoring, and stress analysis in robotics and engineering.

111. Explain the significance of gas sensors in robotics for detecting toxic gases, environmental monitoring, and safety assurance in industrial and hazardous environments.

Answer: Gas sensors are significant in robotics for detecting toxic gases, environmental monitoring, and safety assurance in industrial and hazardous environments. They detect specific gases, ensuring worker safety, pollution control, and environmental protection in applications where gas detection is critical.

112. How do robots utilize angle sensors (potentiometers or encoders) for precise angle measurement, joint positioning, and feedback control in robotic arms and articulated systems?

Answer: Robots utilize angle sensors such as potentiometers or encoders for precise angle measurement, joint positioning, and feedback control in robotic arms and articulated systems. These sensors provide accurate angle data, ensuring precise control, motion planning, and feedback in robotic manipulators and joints.

113. What is the role of pressure transmitters in robotics for monitoring fluid pressure, process control, and hydraulic system management in industrial automation?

Answer: Pressure transmitters play a vital role in robotics for monitoring fluid pressure, process control, and hydraulic system management in industrial automation. They convert fluid pressure into electrical signals, ensuring accurate pressure control, process optimization, and reliable hydraulic system operation in manufacturing and robotics.

114. How can robots employ spectrophotometers for color analysis, quality control, and chemical analysis in applications like food processing and pharmaceuticals?

Answer: Robots can employ spectrophotometers for color analysis, quality control, and chemical analysis in applications like food processing and pharmaceuticals. Spectrophotometers measure absorbance and spectral data, ensuring accurate color analysis, quality assessment, and chemical identification in manufacturing and research.

115. Explain the significance of methane detectors in robotics for methane gas detection, safety monitoring, and environmental protection in industries like oil and gas.

Answer: Methane detectors are significant in robotics for methane gas detection, safety monitoring, and environmental protection in industries like oil and gas. They detect methane gas levels, ensuring worker safety, leak detection, and environmental conservation in applications where methane detection is crucial.

116. What are the advantages of using water level sensors in robotics for water level measurement, flood monitoring, and water resource management?

Answer: Water level sensors offer advantages in robotics for water level measurement, flood monitoring, and water resource management. They measure water levels accurately, enabling flood prediction, water resource management, and efficient water control in environmental and infrastructure applications.

117. How do robots employ temperature transmitters for temperature measurement, process control, and thermal management in industrial and manufacturing processes?

Answer: Robots employ temperature transmitters for temperature measurement, process control, and thermal management in industrial and manufacturing processes. Temperature transmitters convert temperature into electrical signals, ensuring precise temperature control, process optimization, and thermal management in various applications.

118. What is the role of air velocity sensors in robotics for measuring air flow rates, ventilation control, and indoor air quality assessment in HVAC systems?

Answer: Air velocity sensors play a role in robotics for measuring air flow rates, ventilation control, and indoor air quality assessment in HVAC systems. They measure air velocity and flow, ensuring efficient ventilation, air quality management, and energy-saving control in heating, ventilation, and air conditioning.

119. How can robots utilize light-to-voltage sensors (LTV) for light intensity measurement, optical sensing, and automatic lighting control in smart homes and buildings?

Answer: Robots can utilize light-to-voltage sensors (LTV) for light intensity measurement, optical sensing, and automatic lighting control in smart homes and buildings. LTV sensors convert light intensity into voltage signals, enabling light control, optical sensing, and energy-efficient lighting management in home automation and building systems.

120. Explain the significance of capacitance sensors in robotics for non-contact distance measurement, object detection, and proximity sensing in automation and robotics.

Answer: Capacitance sensors are significant in robotics for non-contact distance measurement, object detection, and proximity sensing in automation and robotics. They detect changes in capacitance caused by nearby objects, enabling precise distance measurement, object detection, and proximity sensing in various automation and robotic applications.

121. What are the applications of nuclear radiation detectors in robotics for radiation monitoring, nuclear facility inspection, and safety assurance in nuclear environments?

Answer: Nuclear radiation detectors have applications in robotics for radiation monitoring, nuclear facility inspection, and safety assurance in nuclear environments. They detect ionizing radiation levels, ensuring safety, environmental protection, and reliable operation in nuclear-related applications.

122. How can robots employ oxygen analyzers for oxygen concentration measurement, gas analysis, and oxygen control in medical, industrial, and laboratory settings?

Answer: Robots can employ oxygen analyzers for oxygen concentration measurement, gas analysis, and oxygen control in medical, industrial, and laboratory settings. These analyzers measure oxygen levels accurately, ensuring oxygen therapy control, gas analysis, and precise oxygen management in various applications.

123. Explain the role of turbidity sensors in robotics for water quality assessment, sediment monitoring, and environmental research in aquatic ecosystems.

Answer: Turbidity sensors play a critical role in robotics for water quality assessment, sediment monitoring, and environmental research in aquatic ecosystems. They measure water turbidity, providing data for water quality assessment, sediment monitoring, and ecological research in aquatic environments.

124. What is the significance of load cell amplifiers in robotics for amplifying and conditioning load cell signals, ensuring accurate force and weight measurements?

Answer: Load cell amplifiers are significant in robotics for amplifying and conditioning load cell signals, ensuring accurate force and weight measurements. They enhance signal quality, enabling precise force and weight measurement, process control, and material testing in various robotic and industrial applications.

125. How do robots employ humidity transmitters for humidity measurement, climate control, and environmental monitoring in greenhouses and controlled environments?

Answer: Robots employ humidity transmitters for humidity measurement, climate control, and environmental monitoring in greenhouses and controlled environments. These transmitters measure humidity levels accurately, ensuring precise climate control, plant growth optimization, and environmental monitoring in agriculture and controlled atmospheres.

126. What are the applications of infrared thermopile sensors in robotics for non-contact temperature measurement, thermal imaging, and temperature control in industrial and medical applications?

Answer: Infrared thermopile sensors find applications in robotics for non-contact temperature measurement, thermal imaging, and temperature control in industrial and medical applications. They detect infrared radiation, enabling accurate temperature measurement, thermal imaging, and temperature control in various robotic and industrial processes.

127. How can robots benefit from using angular rate sensors (gyroscopes) for angular velocity measurement, orientation tracking, and motion stabilization in applications like drones and navigation systems?

Answer: Robots benefit from using angular rate sensors (gyroscopes) for angular velocity measurement, orientation tracking, and motion stabilization in applications like drones and navigation systems. Gyroscopes measure angular velocity accurately, ensuring stable flight, precise navigation, and responsive motion control in aerial robotics and navigation systems.

128. Explain the significance of color sensors (RGB sensors) in robotics for color detection, image processing, and color-based object recognition in automation and vision systems.

Answer: Color sensors, or RGB sensors, are significant in robotics for color detection, image processing, and color-based object recognition in automation and vision systems. They capture and analyze color information, enabling robots to identify objects, categorize them by color, and perform color-based tasks in various automation and vision applications.

129. What is the role of laser line scanners in robotics for 3D scanning, object profiling, and obstacle detection in autonomous vehicles and industrial automation?

Answer: Laser line scanners play a role in robotics for 3D scanning, object profiling, and obstacle detection in autonomous vehicles and industrial automation. They project laser lines and capture their reflections, enabling 3D scanning, object profiling, and precise obstacle detection in various applications.

130. How do robots employ sound level meters for sound intensity measurement, noise monitoring, and acoustic analysis in industrial, environmental, and audio applications?

Answer: Robots employ sound level meters for sound intensity measurement, noise monitoring, and acoustic analysis in industrial, environmental, and audio applications. Sound level meters measure sound levels accurately, ensuring noise control, environmental monitoring, and acoustic analysis in various applications.

131. What are the applications of laser rangefinders in robotics for distance measurement, mapping, and 3D modeling in fields like agriculture, forestry, and surveying?

Answer: Laser rangefinders have applications in robotics for distance measurement, mapping, and 3D modeling in fields like agriculture, forestry, and surveying. They emit laser beams to measure distances accurately, enabling precise mapping, terrain analysis, and 3D modeling for applications in these industries.

132. How can robots utilize humidity and temperature data loggers for long-term environmental monitoring, data recording, and research in climate science and ecology?

Answer: Robots can utilize humidity and temperature data loggers for long-term environmental monitoring, data recording, and research in climate science and ecology. These data loggers record humidity and temperature data over extended periods, providing valuable insights for climate research, ecological studies, and environmental monitoring.

133. Explain the role of pyroelectric infrared (PIR) sensors in robotics for motion detection, presence sensing, and security applications in smart homes and buildings.

Answer: Pyroelectric infrared (PIR) sensors play a role in robotics for motion detection, presence sensing, and security applications in smart homes and buildings. PIR sensors detect changes in infrared radiation caused by motion, enabling motion detection, presence sensing, and security alerts, enhancing home automation and security systems.

134. What are the advantages of using pressure transducers with digital outputs (e.g., I2C or SPI) in robotics for pressuremeasurement, data acquisition, and integration into digital systems?

Answer: Pressure transducers with digital outputs, such as I2C or SPI interfaces, offer advantages in robotics for pressure measurement, data acquisition, and integration into digital systems. They provide digital data for easy integration, precise pressure measurement, and data acquisition in robotics and digital control systems.

135. How do robots employ ultrasonic flow meters for measuring fluid flow rates, flow monitoring, and process control in industries like water treatment and oil and gas?

Answer: Robots employ ultrasonic flow meters for measuring fluid flow rates, flow monitoring, and process control in industries like water treatment and oil and gas. These flow meters use ultrasonic waves to measure flow non-invasively, ensuring accurate flow measurement, flow monitoring, and process control in various applications.

136. What is the significance of CO2 gas analyzers in robotics for carbon dioxide measurement, environmental monitoring, and climate research in greenhouse gas analysis?

Answer: CO2 gas analyzers are significant in robotics for carbon dioxide measurement, environmental monitoring, and climate research in greenhouse gas analysis. They measure CO2 levels accurately, providing essential data for environmental monitoring, climate research, and greenhouse gas analysis in various applications.

137. How can robots benefit from using thermal imagers for temperature profiling, thermal analysis, and anomaly detection in applications like industrial maintenance and building inspections?

Answer: Robots benefit from using thermal imagers for temperature profiling, thermal analysis, and anomaly detection in applications like industrial maintenance and building inspections. Thermal imagers capture thermal images, enabling temperature analysis, anomaly detection, and predictive maintenance in various industrial and inspection tasks.

138. Explain the role of linear displacement sensors (LVDTs) in robotics for linear position measurement, feedback control, and precision positioning in automation and manufacturing.

Answer: Linear displacement sensors, or LVDTs, play a role in robotics for linear position measurement, feedback control, and precision positioning in automation and manufacturing. LVDTs measure linear displacement accurately, providing feedback for precise control, motion planning, and positioning in robotics and automated systems.

139. What are the applications of turboprop airspeed sensors in robotics for airspeed measurement, navigation, and flight control in unmanned aerial vehicles (UAVs) and drones?

Answer: Turboprop airspeed sensors have applications in robotics for airspeed measurement, navigation, and flight control in unmanned aerial vehicles (UAVs) and drones. They measure airspeed accurately, ensuring safe and precise flight control, navigation, and performance monitoring in aerial robotics.

140. How do robots employ flame detectors for fire detection, safety monitoring, and flame control in applications like industrial processes and fire protection systems?

Answer: Robots employ flame detectors for fire detection, safety monitoring, and flame control in applications like industrial processes and fire protection systems. Flame detectors detect the presence of flames, ensuring fire safety, safety monitoring, and fire protection in environments where fire detection is critical.

141. What is the significance of laser power meters in robotics for laser power measurement, laser system calibration, and laser safety in laser processing and research?

Answer: Laser power meters are significant in robotics for laser power measurement, laser system calibration, and laser safety in laser processing and research. They measure laser power accurately, ensuring laser system calibration, safety compliance, and precise laser power control in various laser applications.

142. How can robots utilize load cell amplifiers for amplifying and conditioning load cell signals, ensuring accurate force and weight measurements in various applications?

Answer: Robots can utilize load cell amplifiers for amplifying and conditioning load cell signals, ensuring accurate force and weight measurements in various applications. Load cell amplifiers enhance signal quality, enabling precise force and weight measurement, process control, and material testing in robotics and industrial processes.

143. Explain the role of infrared gas sensors in robotics for gas detection, emissions monitoring, and safety assurance in industries like manufacturing and environmental protection.

Answer: Infrared gas sensors play a role in robotics for gas detection, emissions monitoring, and safety assurance in industries like manufacturing and environmental protection. They detect specific gases based on their infrared absorption, ensuring worker safety, emissions control, and environmental protection in various applications.

144. What are the applications of thermal conductivity sensors in robotics for gas concentration measurement, gas analysis, and environmental monitoring in laboratories and industrial settings?

Answer: Thermal conductivity sensors have applications in robotics for gas concentration measurement, gas analysis, and environmental monitoring in laboratories and industrial settings. They measure thermal conductivity of gases, providing data for gas analysis, concentration measurement, and environmental monitoring in various applications.

145. How do robots employ vibration sensors for vibration monitoring, machine condition assessment, and predictive maintenance in industrial machinery and equipment?

Answer: Robots employ vibration sensors for vibration monitoring, machine condition assessment, and predictive maintenance in industrial machinery and equipment. Vibration sensors detect vibrations and provide data for machine health assessment, predictive maintenance, and equipment reliability improvement in various industrial applications.

146. What is the significance of photoelectric sensors in robotics for object detection, counting, and automation in industries like manufacturing and logistics?

Answer: Photoelectric sensors are significant in robotics for object detection, counting, and automation in industries like manufacturing and logistics. They use light beams to detect objects, ensuring accurate object detection, counting, and automation control in various applications.

147. How can robots utilize angle encoders for precise angle measurement, angular positioning, and feedback control in robotic joints, rotary stages, and motion systems?
Answer: Robots can utilize angle encoders for precise angle measurement, angular positioning, and feedback control in robotic joints, rotary stages, and motion systems. Angle encoders provide accurate angle data, ensuring precise control, motion planning, and feedback in various robotic systems.
148. Explain the role of ultraviolet (UV) photodiodes in robotics for UV radiation measurement, UV detection, and UV-sensitive applications in scientific experiments and environmental monitoring.
Answer: Ultraviolet (UV) photodiodes play a role in robotics for UV radiation measurement, UV detection, and UV-sensitive applications in scientific experiments and environmental monitoring. UV photodiodes detect UV radiation, enabling precise UV measurement, UV detection, and monitoring in robotics and scientific research.
149. What are the applications of torque transducers in robotics for torque measurement, load monitoring, and material testing in applications like robotic joints and torque wrenches?
Answer: Torque transducers have applications in robotics for torque measurement, load monitoring, and material testing in applications like robotic joints and torque wrenches. They measure torque accurately, providing data for load monitoring, material testing, and torque control in various robotic and industrial applications.
150. How can robots benefit from using laser beam profilers for laser beam characterization, beam quality analysis, and laser system optimization in laser processing and research?
Answer: Robots can benefit from using laser beam profilers for laser beam characterization, beam quality analysis, and laser system optimization in laser processing and research. Laser beam profilers capture detailed laser beam data, enabling beam quality analysis, system optimization, and precise laser processing in various laser applications.
151. Explain the significance of air quality sensors in robotics for indoor air quality assessment, pollution monitoring, and HVAC control in smart homes and buildings.
Answer: Air quality sensors are significant in robotics for indoor air quality assessment, pollution monitoring, and HVAC control in smart homes and buildings. They measure air quality parameters, ensuring indoor air quality assessment, pollution control, and energy-efficient HVAC systems in home automation and building management.
152. How do robots employ laser displacement sensors (LDS) for non-contact distance measurement, surface profiling, and 3D scanning in applications like robotics and industrial automation?
Answer: Robots employ laser displacement sensors (LDS) for non-contact distance measurement, surface profiling, and 3D scanning in applications like robotics and industrial automation. LDS devices use laser beams to measure distances and profiles, enabling precise measurement, surface analysis, and 3D scanning in various applications.
153. What is the role of gas chromatography (GC) sensors in robotics for chemical analysis, gas separation, and compound detection in research and industrial laboratories?
Answer: Gas chromatography (GC) sensors play a role in robotics for chemical analysis, gas separation, and compound detection in research and industrial laboratories. They separate and analyze complex mixtures of gases and compounds, providing data for chemical analysis, gas separation, and compound detection in various applications.
154. What are the applications of fiber optic sensors in robotics for strain measurement, temperature sensing, and structural health monitoring in civil engineering and aerospace?
Answer: Fiber optic sensors have applications in robotics for strain measurement, temperature sensing, and structural health monitoring in civil engineering and aerospace. They use optical fibers to measure strain and temperature, ensuring accurate data for structural health assessment, temperature monitoring, and aerospace applications.
155. How can robots utilize pH sensors for pH measurement, acidity control, and chemical analysis in industries like pharmaceuticals, food processing, and water treatment?
Answer: Robots can utilize pH sensors for pH measurement, acidity control, and chemical analysis in industries like pharmaceuticals, food processing, and water treatment. PH sensors measure pH levels accurately, ensuring pH control, quality assurance, and chemical analysis in various industrial applications.
156. What are the applications of laser diode modules in robotics for laser projection, alignment, and ranging in applications like 3D printing and laser cutting?

Answer: Laser diode modules have applications in robotics for laser projection, alignment, and ranging in applications like 3D printing and laser cutting. They emit laser beams for precise alignment, projection, and distance measurement, enhancing the capabilities of robots in applications requiring precision and accuracy.

157. How can robots benefit from using gamma radiation detectors for radiation monitoring, nuclear facility inspection, and safety assurance in nuclear and radiation-related industries?

Answer: Robots can benefit from using gamma radiation detectors for radiation monitoring, nuclear facility inspection, and safety assurance in nuclear and radiation-related industries. These detectors measure gamma radiation levels accurately, ensuring safety, environmental protection, and reliable operation in nuclear environments.

158. Explain the role of position encoders in robotics for position feedback, motion control, and positioning accuracy in CNC machines and automated systems.

Answer: Position encoders play a role in robotics for position feedback, motion control, and positioning accuracy in CNC machines and automated systems. They provide precise position data, ensuring accurate motion control, positioning, and quality assurance in various automated and machining applications.

159. What are the applications of gas concentration sensors in robotics for gas detection, environmental monitoring, and safety assurance in confined spaces and hazardous environments?

Answer: Gas concentration sensors have applications in robotics for gas detection, environmental monitoring, and safety assurance in confined spaces and hazardous environments. They detect specific gas concentrations, ensuring worker safety, environmental protection, and gas detection in applications where gas monitoring is essential.

160. How do robots employ inclinometers (tilt sensors) for angle measurement, platform leveling, and stability control in applications like drones and mobile platforms?

Answer: Robots employ inclinometers, also known as tilt sensors, for angle measurement, platform leveling, and stability control in applications like drones and mobile platforms. Inclinometers measure angles accurately, ensuring stable platform operation, precise angle measurement, and control in various robotic and mobile systems.

161. What is the significance of microwave motion sensors in robotics for human presence detection, occupancy sensing, and automation in smart homes and buildings?

Answer: Microwave motion sensors are significant in robotics for human presence detection, occupancy sensing, and automation in smart homes and buildings. They use microwave signals to detect motion, enabling human presence detection, occupancy sensing, and energy-efficient automation in home and building systems.

162. How can robots utilize optical encoders for rotary position sensing and feedback in applications like robotic joints, motor control, and motion systems?

Answer: Robots can utilize optical encoders for rotary position sensing and feedback in applications like robotic joints, motor control, and motion systems. Optical encoders provide accurate angular position data, ensuring precise control, motion planning, and feedback in various robotic and motion control systems.

163. Explain the role of methane analyzers in robotics for methane gas analysis, environmental monitoring, and greenhouse gas research in agriculture and environmental science.

Answer: Methane analyzers play a role in robotics for methane gas analysis, environmental monitoring, and greenhouse gas research in agriculture and environmental science. They measure methane concentrations accurately, providing data for environmental monitoring, greenhouse gas research, and pollution control in various applications.

164. What are the applications of sound pressure level sensors in robotics for sound intensity measurement, noise monitoring, and acoustic analysis in audio and environmental applications?

Answer: Sound pressure level sensors have applications in robotics for sound intensity measurement, noise monitoring, and acoustic analysis in audio and environmental applications. They measure sound pressure levels accurately, ensuring noise control, acoustic analysis, and environmental monitoring in various applications.

165. How do robots employ load cell indicators for displaying and processing load cell data, ensuring real-time force and weight measurement in material testing and industrial applications?

Answer: Robots employ load cell indicators for displaying and processing load cell data, ensuring real-time force and weight measurement in material testing and industrial applications. Load cell indicators provide a user-friendly interface for load cell data, enabling real-time monitoring, process control, and material testing in various applications.

166. What is the role of ultrasonic level sensors in robotics for liquid level measurement, tank monitoring, and industrial automation in industries like chemical processing and wastewater treatment?

Answer: Ultrasonic level sensors play a role in robotics for liquid level measurement, tank monitoring, and industrial automation in industries like chemical processing and wastewater treatment. They use ultrasonic waves to measure liquid levels accurately, ensuring precise tank monitoring, industrial automation, and process control.

167. How can robots benefit from using spectroradiometers for spectral analysis, light measurement, and optical characterization in scientific research and optical systems development?

Answer: Robots can benefit from using Spectroradiometers for spectral analysis, light measurement, and optical characterization in scientific research and optical systems development. Spectroradiometers capture spectral data, enabling spectral analysis, light measurement, and optical system optimization in various research and optical applications.

168. Explain the significance of particulate matter sensors in robotics for air quality monitoring, pollution control, and health assessment in urban environments and industrial settings.

Answer: Particulate matter sensors are significant in robotics for air quality monitoring, pollution control, and health assessment in urban environments and industrial settings. They detect airborne particulate matter, ensuring air quality monitoring, pollution control, and health assessment in areas where air quality is a concern.

169. What are the applications of potentiostats in robotics for electrochemical analysis, corrosion monitoring, and material testing in industries like automotive and aerospace?

Answer: Potentiostats have applications in robotics for electrochemical analysis, corrosion monitoring, and material testing in industries like automotive and aerospace. They control electrochemical experiments, ensuring accurate analysis, corrosion monitoring, and material testing in various industrial applications.

170. How do robots employ linear variable differential transformers (LVDTs) for linear displacement measurement, position feedback, and precision control in applications like hydraulic systems and robotic actuators?

Answer: Robots employ linear variable differential transformers (LVDTs) for linear displacement measurement, position feedback, and precision control in applications like hydraulic systems and robotic actuators. LVDTs provide accurate linear position data, ensuring precise control, motion planning, and feedback in various applications.

171. What is the role of ozone sensors in robotics for ozone gas detection, environmental monitoring, and ozone generation control in applications like water treatment and air purification?

Answer: Ozone sensors play a role in robotics for ozone gas detection, environmental monitoring, and ozone generation control in applications like water treatment and air purification. They detect ozone gas levels accurately, ensuring ozone gas detection, environmental monitoring, and control in various applications.

172. What are the advantages of using strain gauge amplifiers in robotics for amplifying and conditioning strain gauge signals, ensuring precise strain measurement and structural analysis?

Answer: Strain gauge amplifiers offer advantages in robotics for amplifying and conditioning strain gauge signals, ensuring precise strain measurement and structural analysis. They enhance signal quality, enabling accurate strain measurement, structural analysis, and stress assessment in various robotic and structural applications.

173. How can robots utilize thermal anemometers for measuring air velocity, flow rates, and thermal characterization in applications like HVAC systems and fluid dynamics research?

Answer: Robots can utilize thermal anemometers for measuring air velocity, flow rates, and thermal characterization in applications like HVAC systems and fluid dynamics research. Thermal anemometers measure air velocity based on heat transfer, ensuring accurate air velocity measurement, flow rate analysis, and thermal characterization in various applications.

174. Explain the significance of magnetic field sensors in robotics for magnetic field measurement, magnetic field mapping, and navigation in autonomous vehicles and robotics.

Answer: Magnetic field sensors are significant in robotics for magnetic field measurement, magnetic field mapping, and navigation in autonomous vehicles and robotics. They measure magnetic fields, enabling magnetic field mapping, navigation, and positioning in various autonomous systems and robotic applications.

175. What are the applications of force-sensitive resistors (FSRs) in robotics for touch sensing, human-robot interaction, and robotic grippers in collaborative robotics?

Answer: Force-sensitive resistors (FSRs) have applications in robotics for touch sensing, human-robot interaction, and robotic grippers in collaborative robotics. FSRs detect applied force or touch, enabling touch sensing, interaction, and safe operation in collaborative robotic environments.

176. How do robots employ laser safety curtains for laser safety, hazard mitigation, and personnel protection in laser processing and industrial applications?

Answer: Robots employ laser safety curtains for laser safety, hazard mitigation, and personnel protection in laser processing and industrial applications. These curtains block laser beams and ensure safety, hazard mitigation, and personnel protection in environments where laser processing is conducted.

177. What is the role of position transmitters in robotics for linear and rotary position measurement, feedback control, and precision positioning in industrial automation and robotics?

Answer: Position transmitters play a role in robotics for linear and rotary position measurement, feedback control, and precision positioning in industrial automation and robotics. They provide accurate position data, ensuring precise control, motion planning, and feedback in various robotic and automation systems.

178. What are the advantages of using piezoelectric sensors in robotics for vibration detection, dynamic force measurement, and structural health monitoring in aerospace and manufacturing?

Answer: Piezoelectric sensors offer advantages in robotics for vibration detection, dynamic force measurement, and structural health monitoring in aerospace and manufacturing. They convert mechanical vibrations into electrical signals, enabling vibration detection, dynamic force measurement, and structural health assessment in various applications.

179. How can robots benefit from using tilt-compensated compass sensors for accurate orientation sensing, heading determination, and navigation in autonomous systems and robotics?

Answer: Robots can benefit from using tilt-compensated compass sensors for accurate orientation sensing, heading determination, and navigation in autonomous systems and robotics. These sensors compensate for tilt, ensuring accurate orientation data, heading determination, and precise navigation in various robotic and autonomous applications.

180. Explain the significance of vacuum sensors in robotics for vacuum level measurement, process control, and vacuum system management in semiconductor manufacturing and vacuum technology.

Answer: Vacuum sensors are significant in robotics for vacuum level measurement, process control, and vacuum system management in semiconductor manufacturing and vacuum technology. They measure vacuum levels accurately, ensuring precise process control, vacuum system management, and quality assurance in various vacuum-related applications.

181. What are the applications of laser displacement sensors in robotics for distance measurement, object detection, and precision positioning in automation and manufacturing?

Answer: Laser displacement sensors find applications in robotics for distance measurement, object detection, and precision positioning in automation and manufacturing. They emit laser beams to measure distances with high accuracy, enabling precise object detection, distance measurement, and positioning control in various automated processes and manufacturing systems.

182. How can robots utilize color recognition sensors for color-based object sorting, quality control, and material handling in industrial automation and logistics?

Answer: Robots can utilize color recognition sensors for color-based object sorting, quality control, and material handling in industrial automation and logistics. These sensors detect and recognize colors, enabling robots to sort objects based on color, perform quality checks, and handle materials efficiently in manufacturing and logistics operations.

183. Explain the role of spectrophotometers in robotics for spectral analysis, color measurement, and material characterization in industries like textiles and cosmetics.
Answer: Spectrophotometers play a role in robotics for spectral analysis, color measurement, and material characterization in industries like textiles and cosmetics. They measure the absorption and reflection of light across different wavelengths, providing data for color measurement, material analysis, and quality control in various industrial applications.
184. What are the applications of radar sensors in robotics for object detection, collision avoidance, and autonomous navigation in autonomous vehicles and drones?
Answer: Radar sensors have applications in robotics for object detection, collision avoidance, and autonomous navigation in autonomous vehicles and drones. They emit radio waves and detect reflections, enabling accurate object detection, collision avoidance, and autonomous navigation in various autonomous systems and robotics.
185. How do robots employ proximity sensors for detecting nearby objects, obstacle avoidance, and safety assurance in applications like autonomous mobile robots and industrial automation?
Answer: Robots employ proximity sensors for detecting nearby objects, obstacle avoidance, and safety assurance in applications like autonomous mobile robots and industrial automation. Proximity sensors use various technologies (inductive, capacitive, ultrasonic, etc.) to detect objects in their vicinity, ensuring safe and efficient operation in diverse robotic and industrial environments.
186. What is the significance of gas flow meters in robotics for measuring gas flow rates, gas consumption monitoring, and process control in industries like healthcare and gas distribution?
Answer: Gas flow meters are significant in robotics for measuring gas flow rates, gas consumption monitoring, and process control in industries like healthcare and gas distribution. They measure the flow of gases accurately, enabling precise gas consumption monitoring, process control, and quality assurance in various applications.
187. How can robots benefit from using laser range finders for distance measurement, mapping, and navigation in applications like autonomous vehicles and robotic mapping?
Answer: Robots can benefit from using laser range finders for distance measurement, mapping, and navigation in applications like autonomous vehicles and robotic mapping. Laser range finders emit laser beams and measure the time it takes for the beams to bounce back, enabling accurate distance measurement, mapping, and navigation in various robotic and autonomous systems.
188. Explain the role of radiation detectors in robotics for nuclear facility inspection, radiation monitoring, and safety assurance in nuclear and radioactive environments.
Answer: Radiation detectors play a role in robotics for nuclear facility inspection, radiation monitoring, and safety assurance in nuclear and radioactive environments. They detect ionizing radiation levels, ensuring safety, environmental protection, and reliable operation in nuclear-related applications.
189. What are the applications of torque sensors in robotics for torque measurement, robotic joint control, and force feedback in applications like industrial robots and exoskeletons?
Answer: Torque sensors have applications in robotics for torque measurement, robotic joint control, and force feedback in applications like industrial robots and exoskeletons. They measure torque accurately, providing data for precise joint control, force feedback, and user assistance in various robotic and exoskeleton systems.
190. How do robots employ photoionization detectors (PIDs) for volatile organic compound (VOC) detection, environmental monitoring, and safety in industries like chemical manufacturing and environmental protection?
Answer: Robots employ photoionization detectors (PIDs) for volatile organic compound (VOC) detection, environmental monitoring, and safety in industries like chemical manufacturing and environmental protection. PIDs ionize VOCs in the air, enabling VOC detection, environmental monitoring, and safety assurance in various industrial and environmental applications.

191. What is the role of laser power sensors in robotics for laser power measurement, laser system calibration, and laser safety in laser processing and research?

Answer: Laser power sensors play a role in robotics for laser power measurement, laser system calibration, and laser safety in laser processing and research. They measure laser power accurately, ensuring laser system calibration, safety compliance, and precise laser power control in various laser applications.

192. How can robots utilize strain gauges for measuring mechanical strain, stress analysis, and structural health monitoring in applications like robotics and civil engineering?

Answer: Robots can utilize strain gauges for measuring mechanical strain, stress analysis, and structural health monitoring in applications like robotics and civil engineering. Strain gauges change resistance in response to mechanical strain, providing data for stress analysis, structural health monitoring, and performance assessment in various applications.

193. Explain the significance of laser distance sensors in robotics for non-contact distance measurement, object profiling, and obstacle detection in automation and robotics.

Answer: Laser distance sensors are significant in robotics for non-contact distance measurement, object profiling, and obstacle detection in automation and robotics. They use laser beams to measure distances and profiles, enabling precise distance measurement, object profiling, and obstacle detection in various automation and robotic applications.

194. What are the applications of oxygen sensors in robotics for oxygen concentration measurement, gas analysis, and oxygen control in applications like medical devices and environmental monitoring?

Answer: Oxygen sensors have applications in robotics for oxygen concentration measurement, gas analysis, and oxygen control in applications like medical devices and environmental monitoring. They measure oxygen levels accurately, ensuring oxygen therapy control, gas analysis, and precise oxygen management in various applications.

195. How do robots employ color sensors (RGB sensors) for color detection, object recognition, and color-based automation in industries like packaging and quality control?

Answer: Robots employ color sensors, or RGB sensors, for color detection, object recognition, and color-based automation in industries like packaging and quality control. They capture and analyze color information, enabling robots to identify objects, categorize them by color, and perform color-based tasks in various automation and quality control applications.

196. What is the role of ultrasonic level sensors in robotics for liquid level measurement, tank monitoring, and industrial automation in industries like food and beverage processing?

Answer: Ultrasonic level sensors play a role in robotics for liquid level measurement, tank monitoring, and industrial automation in industries like food and beverage processing. They use ultrasonic waves to measure liquid levels accurately, ensuring precise tank monitoring, industrial automation, and process control.

197. What are the advantages of using magnetic encoders in robotics for angle measurement, position feedback, and precision control in applications like robotic arms and CNC machines?

Answer: Magnetic encoders offer advantages in robotics for angle measurement, position feedback, and precision control in applications like robotic arms and CNC machines. They provide accurate angle data, ensuring precise control, motion planning, and feedback in various robotic and machining applications.

198. How can robots benefit from using humidity and temperature sensors for environmental monitoring, climate control, and data logging in applications like greenhouses and smart agriculture?

Answer: Robots can benefit from using humidity and temperature sensors for environmental monitoring, climate control, and data logging in applications like greenhouses and smart agriculture. These sensors measure humidity and temperature accurately, ensuring precise climate control, plant growth optimization, and data logging for smart agriculture.

199. Explain the significance of gas chromatography-mass spectrometry (GC-MS) systems in robotics for chemical analysis, compound identification, and research in forensic science and chemistry.

Answer: Gas chromatography-mass spectrometry (GC-MS) systems are significant in robotics for chemical analysis, compound identification, and research in forensic science and chemistry. They separate and analyze complex mixtures of compounds, providing data for chemical analysis, compound identification, and research in various applications.

200. What are the applications of pressure transmitters in robotics for pressure measurement, data acquisition, and integration into automation systems in industries like oil and gas?

Answer: Pressure transmitters have applications in robotics for pressure measurement, data acquisition, and integration into automation systems in industries like oil and gas. They measure pressure accurately and provide digital data for integration into automation systems, ensuring precise pressure measurement, data acquisition, and process control in various industrial applications.

201. How do robots employ radar level sensors for level measurement, tank gauging, and inventory control in industries like chemical processing and storage facilities?

Answer: Robots employ radar level sensors for level measurement, tank gauging, and inventory control in industries like chemical processing and storage facilities. These sensors use radar waves to measure the level of liquids or solids in tanks, ensuring accurate level measurement, tank gauging, and inventory control in various industrial applications.

202. What is the role of pH transmitters in robotics for pH measurement, pH control, and data transmission in applications like water treatment and chemical manufacturing?

Answer: pH transmitters play a role in robotics for pH measurement, pH control, and data transmission in applications like water treatment and chemical manufacturing. They measure pH levels accurately and transmit data for pH control, quality assurance, and process monitoring in various industrial applications.

203. What are the applications of laser line scanners in robotics for 3D scanning, object modeling, and quality control in industries like automotive design and manufacturing?

Answer: Laser line scanners have applications in robotics for 3D scanning, object modeling, and quality control in industries like automotive design and manufacturing. They project laser lines onto objects and capture 3D data, enabling precise 3D scanning, object modeling, and quality control in various industrial applications.

204. How can robots utilize gas detectors for gas leakage detection, safety monitoring, and emergency response in applications like natural gas distribution and chemical facilities?

Answer: Robots can utilize gas detectors for gas leakage detection, safety monitoring, and emergency response in applications like natural gas distribution and chemical facilities. These detectors detect the presence of hazardous gases, ensuring gas leakage detection, safety monitoring, and timely emergency response in critical environments.

205. Explain the significance of displacement transducers in robotics for linear and angular displacement measurement, position feedback, and precision control in machinery and robotics.

Answer: Displacement transducers are significant in robotics for linear and angular displacement measurement, position feedback, and precision control in machinery and robotics. They provide accurate displacement data, ensuring precise control, motion planning, and feedback in various machinery and robotic systems.

206. What are the applications of load cell amplifiers in robotics for amplifying and conditioning load cell signals, ensuring accurate force and weight measurement in material handling and testing?

Answer: Load cell amplifiers find applications in robotics for amplifying and conditioning load cell signals, ensuring accurate force and weight measurement in material handling and testing. They improve the quality of load cell signals, enabling precise force measurement, material testing, and quality control in various material handling and testing applications.

207. How can robots benefit from using light curtains for presence detection, access control, and safety assurance in applications like industrial automation and conveyor systems?

Answer: Robots can benefit from using light curtains for presence detection, access control, and safety assurance in applications like industrial automation and conveyor systems. Light curtains create a protective barrier and detect intrusions, ensuring safety, access control, and personnel protection in industrial and automation environments.

208. Explain the role of vibration sensors in robotics for vibration analysis, predictive maintenance, and machine health monitoring in manufacturing and industrial machinery.

Answer: Vibration sensors play a role in robotics for vibration analysis, predictive maintenance, and machine health monitoring in manufacturing and industrial machinery. They detect and measure vibrations, providing data for vibration analysis, predictive maintenance, and machine health monitoring to prevent unexpected breakdowns and improve machine efficiency.

209. What are the applications of laser diode drivers in robotics for driving and controlling laser diodes, ensuring stable laser operation in laser-based systems and scientific research?

Answer: Laser diode drivers have applications in robotics for driving and controlling laser diodes, ensuring stable laser operation in laser-based systems and scientific research. They provide precise current and voltage control, enabling stable laser diode operation, beam quality control, and precise laser applications in various fields.

210. How do robots employ humidity transmitters for humidity measurement, data transmission, and environmental monitoring in applications like climate control and HVAC systems?

Answer: Robots employ humidity transmitters for humidity measurement, data transmission, and environmental monitoring in applications like climate control and HVAC systems. Humidity transmitters measure humidity levels accurately and transmit data for climate control, HVAC system optimization, and environmental monitoring in various applications.

211. What is the significance of ultraviolet (UV) sensors in robotics for UV light measurement, UV exposure monitoring, and UV disinfection control in healthcare and water treatment?

Answer: Ultraviolet (UV) sensors are significant in robotics for UV light measurement, UV exposure monitoring, and UV disinfection control in healthcare and water treatment. They measure UV light levels accurately, ensuring UV exposure monitoring, disinfection control, and safety in various applications requiring UV radiation measurement.

212. How can robots utilize force-torque sensors for force and torque measurement, compliance control, and manipulation tasks in collaborative robotics and manufacturing?

Answer: Robots can utilize force-torque sensors for force and torque measurement, compliance control, and manipulation tasks in collaborative robotics and manufacturing. These sensors provide data for precise force and torque measurement, compliance control, and safe interaction between robots and humans in collaborative applications.

213. Explain the role of laser speckle contrast imaging (LSCI) systems in robotics for blood flow imaging, tissue perfusion analysis, and medical diagnostics in healthcare and research.

Answer: Laser speckle contrast imaging (LSCI) systems play a role in robotics for blood flow imaging, tissue perfusion analysis, and medical diagnostics in healthcare and research. They capture and analyze laser speckle patterns, providing data for blood flow imaging, tissue perfusion analysis, and medical diagnostics in various applications.

214. What are the applications of tilt sensors (tilt switches) in robotics for inclination measurement, platform leveling, and attitude control in marine and aerospace systems?

Answer: Tilt sensors, also known as tilt switches, have applications in robotics for inclination measurement, platform leveling, and attitude control in marine and aerospace systems. They detect changes in inclination and provide data for platform leveling, attitude control, and navigation in various marine and aerospace applications.

215. How do robots employ CO2 sensors for carbon dioxide (CO2) measurement, indoor air quality monitoring, and ventilation control in smart buildings and HVAC systems?

Answer: Robots employ CO2 sensors for carbon dioxide (CO2) measurement, indoor air quality monitoring, and ventilation control in smart buildings and HVAC systems. CO2 sensors measure CO2 levels accurately, ensuring indoor air quality monitoring, energy-efficient ventilation control, and comfort in smart building environments.

216. What is the role of current transmitters in robotics for current measurement, data transmission, and motor control in industrial automation and electrical systems?

Answer: Current transmitters play a role in robotics for current measurement, data transmission, and motor control in industrial automation and electrical systems. They measure electrical current accurately and transmit data for motor control, process monitoring, and quality assurance in various industrial and electrical applications.

217. What are the advantages of using thermal cameras in robotics for thermal imaging, temperature measurement, and anomaly detection in industries like predictive maintenance and security?

Answer: Thermal cameras offer advantages in robotics for thermal imaging, temperature measurement, and anomaly detection in industries like predictive maintenance and security. They capture thermal images, enabling temperature measurement, anomaly detection, and predictive maintenance in various industrial and security applications.

218. How can robots benefit from using laser safety goggles for laser eye protection, safety compliance, and personnel safety in laser processing and research?

Answer: Robots can benefit from using laser safety goggles for laser eye protection, safety compliance, and personnel safety in laser processing and research. These goggles block harmful laser wavelengths, ensuring laser eye protection, safety compliance, and personnel safety in environments where lasers are used.

219. Explain the significance of load cell indicators in robotics for displaying and processing load cell data, ensuring real-time force and weight measurement in material testing and industrial applications.

Answer: Load cell indicators are significant in robotics for displaying and processing load cell data, ensuring real-time force and weight measurement in material testing and industrial applications. They provide a user-friendly interface for load cell data, enabling real-time monitoring, process control, and material testing in various applications.

220. What are the applications of colorimeters in robotics for color analysis, color matching, and quality control in industries like printing and textiles?

Answer: Colorimeters have applications in robotics for color analysis, color matching, and quality control in industries like printing and textiles. They measure and analyze color properties, ensuring color matching, quality control, and color consistency in various industrial applications.

221. How do robots employ angular rate sensors (gyroscopes) for angular velocity measurement, orientation sensing, and stabilization in aerospace and autonomous navigation?

Answer: Robots employ angular rate sensors, or gyroscopes, for angular velocity measurement, orientation sensing, and stabilization in aerospace and autonomous navigation. Gyroscopes measure angular rates accurately, providing data for orientation sensing, stabilization, and precise navigation in various aerospace and autonomous systems.

222. What is the role of laser power meters in robotics for laser power measurement, laser system calibration, and laser safety in laser processing and research?

Answer: Laser power meters play a role in robotics for laser power measurement, laser system calibration, and laser safety in laser processing and research. They measure laser power accurately, ensuring laser system calibration, safety compliance, and precise laser power control in various laser applications.

223. How can robots utilize distance measurement sensors for obstacle detection, navigation, and collision avoidance in applications like autonomous vehicles and drones?

Answer: Robots can utilize distance measurement sensors for obstacle detection, navigation, and collision avoidance in applications like autonomous vehicles and drones. These sensors measure distances accurately, enabling obstacle detection, navigation planning, and collision avoidance in various autonomous systems and robotic applications.

224. Explain the significance of light intensity sensors in robotics for light measurement, illumination control, and energy efficiency in smart lighting systems and agriculture.

Answer: Light intensity sensors are significant in robotics for light measurement, illumination control, and energy efficiency in smart lighting systems and agriculture. They measure light intensity accurately, ensuring precise illumination control, energy-efficient lighting, and plant growth optimization in various applications.

225. What are the applications of ultrasonic distance sensors in robotics for distance measurement, object detection, and navigation in applications like autonomous robots and drones?

Answer: Ultrasonic distance sensors have applications in robotics for distance measurement, object detection, and navigation in applications like autonomous robots and drones. They emit ultrasonic waves and measure the time it takes for the waves to bounce back, enabling accurate distance measurement, object detection, and navigation in various robotic and autonomous systems.

226. How do robots employ gas pressure sensors for pressure measurement, leak detection, and pressure control in applications like gas pipelines and industrial equipment?

Answer: Robots employ gas pressure sensors for pressure measurement, leak detection, and pressure control in applications like gas pipelines and industrial equipment. These sensors measure gas pressure accurately, ensuring pressure measurement, leak detection, and pressure control in various industrial and gas-related applications.

227. What is the role of spectrophotometric sensors in robotics for color measurement, spectral analysis, and material characterization in industries like automotive and printing?

Answer: Spectrophotometric sensors play a role in robotics for color measurement, spectral analysis, and material characterization in industries like automotive and printing. They capture spectral data, enabling color measurement, spectral analysis, and material characterization for quality control and process optimization in various applications.

228. How can robots benefit from using laser displacement sensors for distance measurement, object profiling, and position control in applications like robotics and manufacturing?

Answer: Robots can benefit from using laser displacement sensors for distance measurement, object profiling, and position control in applications like robotics and manufacturing. These sensors emit laser beams and measure distances and profiles with high accuracy, enabling precise distance measurement, object profiling, and position control in various robotic and manufacturing systems.

229. Explain the significance of gas flow controllers in robotics for controlling gas flow rates, gas consumption optimization, and process control in industries like semiconductor manufacturing.

Answer: Gas flow controllers are significant in robotics for controlling gas flow rates, gas consumption optimization, and process control in industries like semiconductor manufacturing. They regulate gas flow rates precisely, ensuring gas consumption optimization, process control, and quality assurance in various industrial applications.

ONLINE RESOURCES

Practice Paper Link: https://forms.gle/DHAdFPzZXUFQDAxn8

QR Code:

Video Content Link:

https://www.youtube.com/playlist?list=PLzlh_31VWEFrpZQQQsnvJfvqtf77rxOJJ

QR Code:

Chapter 6: Robot Actuators

1. What is the role of actuators in robotics?

Answer: Actuators in robotics are essential components responsible for converting electrical signals into mechanical motion. They enable robots to perform tasks by moving their limbs, grippers, or other parts.

2. Can you explain the types of actuators commonly used in robotics?

Answer: Common types of actuators in robotics include electric motors (DC, AC), pneumatic actuators, hydraulic actuators, and piezoelectric actuators. Each type has unique characteristics suited for various applications.

3. How do electric motors function as actuators in robots?

Answer: Electric motors, such as DC and AC motors, use electromagnetic fields to generate rotary motion in robotics. They are widely used for tasks requiring precision and control.

4. What are the advantages of pneumatic actuators in robotics?

Answer: Pneumatic actuators offer rapid motion and are lightweight, making them suitable for applications that require high-speed movements and minimal weight on robot arms.

5. In what situations are hydraulic actuators preferred in robotic systems?

Answer: Hydraulic actuators excel in applications demanding high force output, like heavy lifting and industrial automation, due to their ability to generate substantial power.

6. How do piezoelectric actuators contribute to precision in robotics?

Answer: Piezoelectric actuators are known for their nanometer-scale precision and fast response times. They find use in applications such as micro-manipulation and optics positioning.

7. What role do shape memory alloys play as actuators in robotics?

Answer: Shape memory alloys (SMAs) are materials that change shape in response to temperature variations. They are used as actuators in robotics for tasks like shape-changing and morphing structures.

8. Can you explain the concept of actuators with feedback control in robotics?

Answer: Actuators with feedback control incorporate sensors to provide information about their position and performance. This feedback helps robots achieve precise movements and adjust to changing conditions.

9. How do robotic grippers utilize actuators for grasping objects?

Answer: Robotic grippers often employ actuators to open and close, enabling robots to grasp and manipulate objects. These actuators vary in type, such as electric or pneumatic, based on the gripper design.

10. What is the significance of actuator selection in robotic design?

Answer: Choosing the right actuator is crucial as it directly impacts a robot's performance, capabilities, and efficiency. The selection process considers factors like payload, speed, precision, and power source.

11. How do robots use linear actuators for precise linear motion?

Answer: Linear actuators convert rotational motion into linear motion, allowing robots to move along a straight path with precision. They are used in applications such as CNC machines and robotics arms.

12. Can you explain the concept of compliant actuators in robotics?

Answer: Compliant actuators are designed to be flexible, absorbing shocks and adapting to their surroundings. They are valuable in collaborative robotics, where robots work alongside humans.

13. What are the key challenges in designing miniature actuators for micro-robotics?

Answer: Miniature actuators for micro-robotics face challenges related to power efficiency, heat management, and maintaining precision on a small scale.

14. How can robots benefit from the use of rotary actuators?

Answer: Rotary actuators enable robots to rotate components or perform tasks involving rotational motion. They are employed in applications like robotic arms and conveyor systems.

15. What are the considerations for actuator control methods in robotics?

Answer: Actuator control methods include position control, velocity control, and force control. The choice depends on the specific task and the level of control required.

16. How do robots achieve smooth and precise motion using stepper motors as actuators?
Answer: Stepper motors move in discrete steps, allowing precise control over position and speed. They are favored in applications where accuracy is essential, such as 3D printers.
17. What advantages do voice coil actuators offer in robotics applications?
Answer: Voice coil actuators provide fast and accurate linear motion, making them suitable for applications like autofocus cameras and robotics requiring rapid adjustments.
18. Can you explain the role of electroactive polymers (EAPs) as soft actuators in robotics?
Answer: Electroactive polymers (EAPs) are flexible materials that deform when subjected to an electric field. They are used in soft robotics for applications like delicate gripping and bio-inspired locomotion.
19. How do robots employ magnetic actuators for precise and contactless motion?
Answer: Magnetic actuators, such as solenoids and magnetic levitation systems, use magnetic fields to generate motion without physical contact. They find use in precision applications like magnetic bearings.
20. What are the challenges in implementing actuator redundancy in robotic systems?
Answer: Implementing actuator redundancy enhances fault tolerance but also introduces complexity in control algorithms and additional hardware, increasing cost and weight.
21. How do robots ensure safety when using high-powered actuators?
Answer: Safety measures for high-powered actuators include implementing protective barriers, emergency stop systems, and thorough risk assessments to prevent accidents.
22. What role do actuators play in underwater robotics?
Answer: Actuators in underwater robots enable propulsion, buoyancy control, and manipulation of objects in the underwater environment. They are essential for tasks like ocean exploration and maintenance.
23. Can you explain the concept of biomimetic actuators in robotics?
Answer: Biomimetic actuators mimic natural systems and movements. They are used in bio-inspired robots to replicate the mobility and behaviors of animals and organisms.
24. How do robots use shape memory polymer actuators for morphing structures?
Answer: Shape memory polymer actuators change shape in response to temperature variations, allowing robots to alter their structure for various purposes, such as adapting to different tasks.
25. What are the key considerations for actuator maintenance in robotics?
Answer: Actuator maintenance involves regular checks for wear and tear, lubrication, and monitoring of sensors and feedback systems to ensure reliable robot performance.
26. How do robots integrate actuators with control algorithms for autonomous navigation?
Answer: Robots use control algorithms to process sensor data and make decisions about actuator movements, enabling them to navigate autonomously and avoid obstacles.
27. What are the challenges in achieving energy-efficient actuator designs for mobile robots?
Answer: Energy-efficient actuator designs for mobile robots must balance power consumption with performance, often requiring trade-offs and innovations in materials and control strategies.
28. Can you explain the concept of haptic actuators in robotics?
Answer: Haptic actuators provide tactile feedback to users or robots, enhancing the sense of touch. They are used in applications like virtual reality, telemedicine, and surgical robotics.
29. What role do shape memory alloy actuators play in robotics for adaptive structures?
Answer: Shape memory alloy actuators enable robots to change their shape or stiffness, making them adaptable to different environments or tasks. They are used in applications like space exploration and search and rescue.
30. How do robots ensure the reliability of their actuators in critical applications like medical robotics?
Answer: In critical applications like medical robotics, redundancy, rigorous testing, and fail-safe mechanisms are employed to ensure the reliability and safety of actuators, minimizing the risk of system failures.
31. What are the advantages of using rotary vane actuators in robotics?
Answer: Rotary vane actuators offer compact design, high torque output, and precise rotational motion. They find use in applications requiring rotational movement in tight spaces, such as robot joints and valve control.

32. How do pneumatic muscle actuators enhance the flexibility of soft robots?
Answer: Pneumatic muscle actuators, made of flexible materials, mimic human muscles' flexibility and adaptability. They enable soft robots to achieve complex motions, making them suitable for tasks like medical devices and search-and-rescue missions.
33. Can you explain the concept of shape deposition manufacturing (SDM) for creating customized shape memory alloy actuators?
Answer: Shape deposition manufacturing (SDM) is a 3D printing technique that produces custom-shaped components, including shape memory alloy actuators. It allows for precise control and customization of actuator shapes and functionalities.
34. What challenges do robots face when using bi-directional shape memory alloy actuators?
Answer: Bi-directional shape memory alloy actuators have challenges related to managing thermal hysteresis and controlling two-way shape memory effects, necessitating careful control algorithms and thermal management systems.
35. How do hybrid actuators, combining different actuation methods, improve robotics performance?
Answer: Hybrid actuators leverage the strengths of multiple actuation methods, such as combining electric and hydraulic systems, to enhance a robot's efficiency, speed, and power, opening up possibilities for versatile applications.
36. What role do artificial muscles play as biomimetic actuators in robotics?
Answer: Artificial muscles emulate natural muscle structures, providing robots with soft, flexible, and energy-efficient actuation for lifelike movements and human-robot interaction in fields like rehabilitation and prosthetics.
37. How do robots utilize magnetostrictive actuators for precise linear motion and position control?
Answer: Magnetostrictive actuators change shape in response to a magnetic field, enabling precise linear motion and position control in applications like precision manufacturing and robotics.
38. Can you explain the use of cable-driven actuators in exoskeletons and assistive robotics?
Answer: Cable-driven actuators use cables and pulleys to transmit motion, making them lightweight and suitable for exoskeletons and assistive robotics to enhance human mobility and strength.
39. What challenges arise in implementing artificial intelligence (AI) for actuator control in robotics?
Answer: AI-driven actuator control requires advanced algorithms and sensor fusion techniques to adapt to dynamic environments, introducing challenges related to computational complexity and real-time processing.
40. How do robots mitigate the effects of backlash in gear-driven actuators for accurate motion control?
Answer: Robots compensate for backlash, the mechanical play in gear-driven actuators, by employing control algorithms that anticipate and correct for the undesired motion, ensuring accurate and precise movements.
41. What are the benefits of using rotary piezoelectric actuators in optics and photonics applications?
Answer: Rotary piezoelectric actuators offer sub-nanometer precision and rapid response, making them ideal for applications requiring precise optical positioning and beam control, such as laser systems and microscopy.
42. How can robots achieve energy-efficient actuation using flywheel energy storage systems?
Answer: Flywheel energy storage systems store kinetic energy in a rotating flywheel and release it when needed, offering energy-efficient actuation for robots in applications like clean energy storage and high-speed motion.
43. Can you explain the role of contractile gel actuators in soft robotics for bio-inspired locomotion?
Answer: Contractile gel actuators mimic muscle contractions and enable soft robots to achieve lifelike movements, making them suitable for underwater exploration and bio-inspired robots.
44. What considerations are essential when designing redundant actuation systems for robotic manipulators?
Answer: Redundant actuation systems enhance robot flexibility but require careful planning to ensure coordinated motion, avoid singularities, and optimize control strategies in complex environments.
45. How do robots use triboelectric actuators for energy harvesting and self-powered sensing?
Answer: Triboelectric actuators generate electrical energy from friction, enabling robots to harvest energy from their movements and power self-sensing systems for prolonged autonomy.

46. What challenges do robots face in controlling shape memory alloy wire actuators for shape-changing applications?

Answer: Shape memory alloy wire actuators have challenges related to nonlinear behavior and hysteresis, necessitating advanced control algorithms for precise and reliable shape-changing transformations.

47. How can robots ensure the longevity of their actuators in harsh environments like space exploration?

Answer: In space exploration, robots employ radiation-hardened actuators and protective coatings to withstand extreme conditions and maintain actuator performance throughout mission durations.

48. What are the advantages of using smart materials like piezoelectric ceramics as actuators in micro-robotics?

Answer: Piezoelectric ceramics offer micro-robots high precision and quick response times for delicate tasks like micro assembly and medical procedures.

49. How do robots use magnetic levitation (maglev) systems as actuators for contactless and frictionless motion?

Answer: Maglev actuators use magnetic fields to levitate objects, enabling robots to achieve smooth and frictionless motion in applications such as transportation and magnetic bearings.

50. Can you explain the concept of hydrogel-based actuators in soft robotics for bio-inspired locomotion and manipulation?

Answer: Hydrogel-based actuators swell and contract in response to changes in humidity or pH, enabling soft robots to mimic natural movements for applications in medicine and environmental monitoring.

51. How do robots employ thermal actuators in micro-robotics for temperature-responsive actuation and control?

Answer: Thermal actuators use thermal expansion properties to achieve controlled movements in micro-robotics, allowing for precise temperature-responsive actuation.

52. What is the role of compliant actuators in robotic exposits for enhancing human strength and mobility?

Answer: Compliant actuators in robotic exposits provide flexible assistance to augment human movements, improving endurance and reducing muscle fatigue in applications like military and healthcare.

53. How can robots benefit from the use of biomimetic electroactive polymer actuators for agile and energy-efficient locomotion?

Answer: Biomimetic electroactive polymer actuators emulate natural muscle movements, enabling robots to achieve agile and energy-efficient locomotion in applications like bio-inspired robotics and prosthetics.

54. What challenges arise in designing actuators for soft robotics that interact safely with humans in collaborative environments?

Answer: Designing safe actuators for soft robotics in collaborative settings involves addressing issues related to impact force, compliance, and sensing to ensure human-robot interaction is secure and injury-free.

55. How do robots use pneumatic artificial muscles (PAMs) as actuators for human-like motion in humanoid robots?

Answer: PAMs mimic human muscles' contracting and elongating motion, enabling humanoid robots to achieve human-like movements and flexibility in applications like human-robot interaction and entertainment.

56. What are the considerations for actuator redundancy in surgical robotics for minimally invasive procedures?

Answer: Actuator redundancy in surgical robotics enhances precision but requires careful control strategies and safety mechanisms to prevent errors during minimally invasive surgeries.

57. How can robots optimize actuator designs for energy-efficient locomotion in mobile robotics?

Answer: Optimizing actuator designs for energy efficiency involves minimizing friction, reducing weight, and employing advanced control algorithms for mobile robots in applications like autonomous vehicles and drones.

58. Can you explain the role of dielectric elastomer actuators (DEAs) in soft robotics for versatile and deformable structures?
Answer: DEAs are soft and flexible actuators that expand and contract when voltage is applied, enabling soft robots to achieve versatile and deformable structures for applications like wearable technology and soft grippers.
59. How do robots use ionic polymer-metal composites (IPMCs) as actuators for underwater robotics and biomimetic propulsion?
Answer: IPMCs respond to changes in voltage by bending and flexing, making them suitable for underwater robotics and biomimetic propulsion, where their flexibility mimics fish-like motion.
60. What challenges do robots face in controlling dielectric elastomer actuators (DEAs) for precise deformations and shape-changing capabilities?
Answer: DEAs have challenges related to nonlinear behavior and voltage control for precise deformations, requiring advanced control algorithms for shape-changing applications in soft robotics.
61. How can robots employ magnetic shape-memory alloy actuators for reversible shape transformations in space exploration and aerospace applications?
Answer: Magnetic shape-memory alloy actuators utilize magnetic fields to induce reversible shape changes, making them valuable for applications in space exploration, aerospace, and deployable structures where shape transformations are required.
62. What role do hydraulic rotary actuators play in heavy-duty industrial robots and construction machinery?
Answer: Hydraulic rotary actuators offer robust and high-torque rotational motion, making them suitable for heavy-duty industrial robots and construction equipment for tasks like excavating and lifting.
63. How do robots use tendon-driven actuators in bio-inspired robotic limbs for natural and dexterous movements?
Answer: Tendon-driven actuators mimic human muscle-tendon systems, enabling bio-inspired robotic limbs to achieve natural and dexterous movements in applications like prosthetics and humanoid robotics.
64. What challenges arise in the integration of soft pneumatic actuators in wearable robotics for rehabilitation and assistance?
Answer: Integrating soft pneumatic actuators in wearable robotics requires addressing issues related to power efficiency, control, and the design of comfortable interfaces to assist individuals in rehabilitation and daily tasks.
65. How can robots ensure actuator synchronization in multi-joint robotic arms for precision and coordinated motion?
Answer: Synchronizing actuators in multi-joint robotic arms involves using feedback control loops and kinematic models to ensure precise and coordinated motion in applications like manufacturing and robotics.
66. What advantages do magnetostrictive actuators offer in ultrasonic machining for precision material removal?
Answer: Magnetostrictive actuators enable precise ultrasonic vibrations for material removal in applications like ultrasonic machining, enhancing precision and surface finish.
67. How do robots employ helical actuators for extending and retracting linear motion in constrained spaces?
Answer: Helical actuators provide linear motion while maintaining a compact form, making them suitable for tasks in confined spaces where extended reach is required.
68. Can you explain the role of low-friction actuators in reducing energy consumption in mobile robotics?
Answer: Low-friction actuators minimize energy loss due to friction, improving the energy efficiency of mobile robots in applications such as autonomous exploration and surveillance.
69. What challenges do robots face in achieving actuator miniaturization for micro-robotics and biomedical devices?
Answer: Miniaturizing actuators for micro-robotics and biomedical devices poses challenges related to power density, precision, and heat dissipation, necessitating innovative design and manufacturing techniques.

70. How can robots optimize piezoelectric stack actuators for high-force applications like precision machining?

Answer: Optimizing piezoelectric stack actuators involves stacking multiple layers to increase force output and using feedback control for precise positioning in high-force applications.

71. What is the significance of compliance control in pneumatic actuators for safe human-robot collaboration in industrial settings?

Answer: Compliance control in pneumatic actuators allows robots to exhibit flexibility and adaptability, ensuring safe interactions with humans in collaborative industrial environments.

72. How do robots utilize thermo-responsive actuators in self-assembling structures for space exploration and autonomous construction?

Answer: Thermo-responsive actuators change shape in response to temperature variations, enabling self-assembling structures for space exploration and autonomous construction of habitats.

73. What role do electrostatic actuators play in micro-robotics for miniature and delicate tasks?

Answer: Electrostatic actuators generate forces by exploiting electrostatic attraction or repulsion, making them suitable for miniature and delicate tasks in micro-robotics, such as micro assembly and micromanipulation.

74. Can you explain the use of water-based hydraulic actuators in soft robots for underwater exploration and aquatic research?

Answer: Water-based hydraulic actuators utilize water pressure for motion, making them ideal for underwater robots in environments where conventional actuators may not perform well.

75. How do robots employ dielectric elastomer stack actuators for soft grippers and adaptive structures?

Answer: Dielectric elastomer stack actuators consist of layers that expand or contract when voltage is applied, making them versatile for soft grippers and adaptive structures in robotics.

76. What are the considerations for implementing rotary voice coil actuators in precision optical systems and camera platforms?

Answer: Implementing rotary voice coil actuators involves addressing factors like speed, precision, and feedback control to ensure optimal performance in precision optical systems and camera platforms.

77. How can robots benefit from using magnetic shape-memory polymer actuators for shape-memory and adaptive structures in aerospace applications?

Answer: Magnetic shape-memory polymer actuators enable shape-memory and adaptive structures in aerospace applications, where their reversible shape changes and lightweight properties are advantageous.

78. Can you explain the role of hybrid hydraulic-electric actuators in mobile construction and agricultural robots?

Answer: Hybrid hydraulic-electric actuators combine the advantages of hydraulic power with electric control, offering high power output and energy efficiency in mobile construction and agricultural robots.

79. What challenges do robots face in controlling piezoelectric bimorph actuators for precision positioning and scanning in microscopy?

Answer: Controlling piezoelectric bimorph actuators for precision positioning and scanning in microscopy requires addressing challenges related to hysteresis and dynamic control for accurate movements.

80. How do robots use carbon nanotube actuators for high-speed and lightweight applications in aerospace and robotics?

Answer: Carbon nanotube actuators leverage the unique properties of carbon nanotubes to achieve high-speed and lightweight actuation in aerospace and robotics, enabling rapid movements and reduced weight.

81. What is the significance of hydraulic hybrid actuators in off-road and heavy-duty industrial vehicles?

Answer: Hydraulic hybrid actuators enhance the efficiency and power of off-road and heavy-duty industrial vehicles by utilizing hydraulic systems to provide additional power for acceleration and deceleration.

82. How can robots achieve energy-efficient actuation using shape memory alloy foam actuators for soft robotics and wearable technology?

Answer: Shape memory alloy foam actuators offer energy-efficient actuation for soft robotics and wearable technology, where their lightweight and adaptable nature contribute to prolonged operation.

83. What role do phase-change materials play in thermally responsive actuators for micro-robotics and medical devices?
Answer: Phase-change materials enable thermally responsive actuators in micro-robotics and medical devices, where their ability to change state with temperature provides precise control and actuation.
84. How do robots employ pneumatic rotary vane actuators for continuous rotary motion in manufacturing and automation?
Answer: Pneumatic rotary vane actuators provide continuous rotary motion in manufacturing and automation applications, offering simplicity and reliability for tasks like valve control and conveyor systems.
85. What are the challenges in implementing biodegradable actuators for environmentally friendly robotics and disposable medical devices?
Answer: Implementing biodegradable actuators in robotics and medical devices involves addressing issues related to material degradation and controlled actuation for eco-friendly and disposable applications.
86. How can robots utilize inflatable actuators for shape-changing and soft locomotion in planetary exploration and search-and-rescue missions?
Answer: Inflatable actuators enable robots to achieve shape-changing and soft locomotion in planetary exploration and search-and-rescue missions, where adaptability to diverse environments is essential.
87. What is the significance of shape memory alloy springs as actuators in aerospace and automotive applications?
Answer: Shape memory alloy springs serve as lightweight and reliable actuators in aerospace and automotive applications, providing precise control and shape-changing capabilities.
88. How do robots employ magnetorheological (MR) actuators for adjustable stiffness and damping in vibration control and haptic feedback systems?
Answer: MR actuators use magnetic fields to control stiffness and damping properties, making them valuable for vibration control, haptic feedback, and adaptive structures in robotics and engineering.
89. What challenges do robots face when using piezoelectric fiber composite actuators for bio-inspired propulsion and biomimetic robots?
Answer: Piezoelectric fiber composite actuators have challenges related to integration and scalability in bio-inspired propulsion and biomimetic robots, requiring careful design and control.
90. How can robots optimize smart fluid actuators for variable stiffness and shape-changing in soft robotics and medical devices?
Answer: Smart fluid actuators utilize fluids that change viscosity with applied voltage, enabling variable stiffness and shape-changing in soft robotics and medical devices for tasks like catheter navigation.
91. What role do modular actuators play in simplifying robotic system design and maintenance?
Answer: Modular actuators simplify robotic system design and maintenance by providing interchangeable components that reduce downtime and allow for efficient upgrades or replacements.
92. How do robots employ shape memory alloy ribbons for actuation in micro-robotics and MEMS devices?
Answer: Shape memory alloy ribbons serve as compact and reliable actuators in micro-robotics and MEMS devices, enabling precise control in miniature applications.
93. What are the advantages of using rotary piezoelectric actuators for beam steering in optical communication and laser systems?
Answer: Rotary piezoelectric actuators offer precise beam steering in optical communication and laser systems, enhancing accuracy and alignment in applications like telecommunications and research.
94. How can robots benefit from the use of biohybrid actuators that combine biological and synthetic components for locomotion and manipulation?
Answer: Biohybrid actuators combine biological and synthetic elements to achieve lifelike locomotion and manipulation in robots for applications like bio-inspired robotics and tissue engineering.

95. What considerations are essential when implementing phase-change actuators for thermal control and shape adjustment in satellites and space missions?
Answer: Implementing phase-change actuators in satellites and space missions requires careful thermal management and control strategies to ensure reliable thermal control and shape adjustment.
96. How do robots utilize magnetic field-driven actuators for non-contact and precise motion in micro-robotics and lab-on-a-chip devices?
Answer: Magnetic field-driven actuators achieve non-contact and precise motion in micro-robotics and lab-on-a-chip devices, enabling manipulation and analysis of tiny samples and components.
97. What is the role of spring-driven actuators in kinetic art and interactive installations?
Answer: Spring-driven actuators provide kinetic motion in artistic installations and interactive exhibits, creating dynamic and engaging experiences for viewers.
98. How can robots optimize hybrid electrostatic-piezoelectric actuators for fine positioning and micro-assembly in electronics manufacturing?
Answer: Hybrid electrostatic-piezoelectric actuators offer fine positioning and micro-assembly capabilities in electronics manufacturing, contributing to precision and efficiency in assembly processes.
99. What challenges do robots face in controlling magnetostrictive micro actuators for microfluidic devices and lab-on-a-chip systems?
Answer: Controlling magnetostrictive micro actuators for microfluidic devices and lab-on-a-chip systems involves addressing challenges related to scalability and integration for precise fluid manipulation.
100. How do robots use ionic liquid-based actuators for responsive motion and adaptable structures in soft robotics and adaptive optics?
Answer: Ionic liquid-based actuators enable responsive motion and adaptable structures in soft robotics and adaptive optics, where their versatility and controllability enhance performance in changing environments.
101. What role do compliant fluidic actuators play in soft robotics for safe human interaction and delicate object handling?
Answer: Compliant fluidic actuators offer safe human-robot interaction and delicate object handling in soft robotics. They utilize fluids to achieve flexible and adaptable motions, ideal for applications like medical devices and collaborative robots.
102. How can robots utilize shape memory polymer foams as actuators for adaptive structures and cushioning in robotics and aerospace?
Answer: Shape memory polymer foams act as actuators for adaptive structures and cushioning in robotics and aerospace. They provide shape-changing capabilities and impact resistance, enhancing safety and functionality.
103. What challenges arise in controlling fiber-reinforced pneumatic actuators for bio-inspired locomotion and biomimetic robots?
Answer: Controlling fiber-reinforced pneumatic actuators for bio-inspired locomotion and biomimetic robots requires addressing issues related to modeling and synchronization to achieve lifelike movements.
104. How do robots employ shape memory alloy coils as actuators for adjustable antennas and deployable structures in space exploration?
Answer: Shape memory alloy coils serve as actuators for adjustable antennas and deployable structures in space exploration, providing precise control and compact solutions for satellite missions.
105. What are the advantages of using eddy current actuators in precision positioning and magnetic levitation systems?
Answer: Eddy current actuators offer precision positioning and non-contact magnetic levitation, making them valuable for applications like semiconductor manufacturing and magnetic bearings.
106. How can robots benefit from the use of ionic polymer-metal composites (IPMCs) as soft actuators for aquatic robots and underwater manipulation?
Answer: IPMCs serve as soft actuators for aquatic robots and underwater manipulation due to their flexibility and underwater compatibility, enabling tasks like ocean exploration and marine research.

107. What is the significance of opt-mechanical actuators in optical instrumentation and laser beam steering?
Answer: Opt-mechanical actuators are critical in optical instrumentation and laser beam steering, providing controlled movements and precise alignment for applications in astronomy and telecommunications.
108. How do robots employ shape memory alloy wires for shape-changing antennas and adaptive structures in aerospace and satellite systems?
Answer: Shape memory alloy wires act as actuators for shape-changing antennas and adaptive structures in aerospace and satellite systems, offering compact and reliable solutions for changing mission requirements.
109. What role do twisted and coiled polymer actuators (TCPAs) play in soft robotics for intricate motions and bio-inspired locomotion?
Answer: TCPAs contribute to intricate motions and bio-inspired locomotion in soft robotics. Their twisted and coiled structure enables versatile movements, making them suitable for tasks like crawling and grasping.
110. How can robots optimize pneumatic artificial muscles (PAMs) for rehabilitation exoskeletons and assistive devices?
Answer: Optimizing PAMs for rehabilitation exoskeletons and assistive devices involves customization and control strategies to enhance mobility and provide support during rehabilitation therapy.
111. What are the considerations for using magnetostrictive thin films as actuators in MEMS devices and micro-robotics?
Answer: Using magnetostrictive thin films as actuators in MEMS devices and micro-robotics requires attention to material properties and microfabrication techniques to achieve precise and responsive actuation.
112. How do robots employ liquid crystal elastomer (LCE) actuators for tunable optics and adaptive camouflage in military and defense applications?
Answer: LCE actuators enable tunable optics and adaptive camouflage in military and defense applications, offering rapid adjustments for optical systems and concealment in changing environments.
113. What challenges do robots face when implementing Ferro fluidic actuators for precise fluid manipulation and micro pump systems?
Answer: Implementing Ferro fluidic actuators for precise fluid manipulation and micro pump systems involves addressing challenges related to magnetic field control and miniaturization for microfluidic applications.
114. How can robots utilize gel-based shape memory actuators for biomedical devices and controlled drug delivery systems?
Answer: Gel-based shape memory actuators serve in biomedical devices and controlled drug delivery systems, providing controlled movement and responsive drug release for therapeutic applications.
115. What is the role of actuator redundancy in advanced humanoid robots and disaster response robots?
Answer: Actuator redundancy enhances advanced humanoid robots and disaster response robots' flexibility and reliability, allowing them to continue functioning even in the presence of actuator failures.
116. How do robots employ adaptive airfoil actuators for wing shape adjustments in unmanned aerial vehicles (UAVs) and drones?
Answer: Adaptive airfoil actuators enable wing shape adjustments in UAVs and drones, optimizing aerodynamics and flight performance for increased efficiency and maneuverability.
117. What advantages do dielectric elastomer generators (DEGs) offer in energy harvesting applications for self-powered sensors and remote monitoring?
Answer: DEGs serve as energy harvesters for self-powered sensors and remote monitoring systems, converting mechanical energy into electrical energy for extended device autonomy.
118. How can robots optimize thermal-responsive shape memory polymer actuators for thermal regulation in space habitats and climate control systems?
Answer: Thermal-responsive shape memory polymer actuators can regulate temperature in space habitats and climate control systems by expanding and contracting in response to temperature changes, ensuring comfort and efficiency.

119. What challenges arise in controlling smart material actuators for adaptive optics systems and wavefront correction in astronomical instruments?

Answer: Controlling smart material actuators for adaptive optics and wavefront correction in astronomical instruments involves addressing challenges related to sensor feedback and real-time adjustments for precise imaging.

120. How do robots employ cable-driven parallel actuators for heavy lifting and manipulation tasks in industrial automation?

Answer: Cable-driven parallel actuators provide heavy lifting and manipulation capabilities in industrial automation, offering high payload capacity and precision for manufacturing processes.

121. What are the advantages of using shape memory alloy springs as actuators in miniature robotics for medical procedures and microsurgery?

Answer: Shape memory alloy springs serve as actuators in miniature robotics for medical procedures and microsurgery, providing compact and precise motion for delicate operations.

122. How can robots benefit from the use of triboelectric Nano generators (TENGs) as actuators for self-powered sensing and energy harvesting in autonomous systems?

Answer: TENGs serve as self-powered sensors and energy harvesters in autonomous systems, utilizing friction-generated electricity to enhance sensing capabilities and extend operation.

123. What is the significance of soft dielectric elastomer actuators (DEAs) in soft robotics for versatile motion and adaptive structures?

Answer: Soft DEAs play a crucial role in soft robotics by providing versatile motion and adaptability, enabling applications like soft grippers and wearable assistive devices.

124. How do robots utilize ionic polymer-metal composites (IPMCs) as actuators for underwater robotics and aquatic exploration?

Answer: IPMCs act as actuators for underwater robotics and aquatic exploration due to their underwater compatibility and ability to mimic fish-like motion for efficient propulsion.

125. What challenges do robots face in controlling pneumatic bellows actuators for versatile and dynamic motions in humanoid robots and exoskeletons?

Answer: Controlling pneumatic bellows actuators in humanoid robots and exoskeletons involves addressing challenges related to real-time control and synchronization for dynamic and versatile movements.

126. How can robots optimize voice coil actuators for rapid and precise motion in micro-optics and Nano manipulation applications?

Answer: Optimizing voice coil actuators involves feedback control and precision tuning for rapid and precise motion in micro-optics and Nano manipulation tasks, enhancing performance and accuracy.

127. What role do hydrogel-based actuators play in soft robotics for bio-inspired locomotion and manipulation in aquatic environments?

Answer: Hydrogel-based actuators contribute to bio-inspired locomotion and manipulation in soft robotics for aquatic environments, providing flexibility and adaptability in underwater exploration.

128. How do robots employ piezoelectric micro actuators for precise and controlled movements in MEMS-based devices and optical switches?

Answer: Piezoelectric micro actuators enable precise and controlled movements in MEMS-based devices and optical switches, enhancing functionality and performance in various applications.

129. What are the considerations for using shape memory alloy wires as actuators in deployable structures for space exploration and satellite deployments?

Answer: Implementing shape memory alloy wires as actuators for deployable structures in space exploration involves considerations of reliability, thermal management, and integration for successful missions.

130. How can robots utilize magnetic levitation (maglev) actuators for contactless and frictionless motion in transportation and magnetic bearings?

Answer: Maglev actuators offer contactless and frictionless motion in transportation and magnetic bearings, reducing wear and enabling smooth and efficient movement in various applications.

131. What role do oscillating water columns serve as wave energy converters and self-powered actuators in marine robotics and renewable energy systems?

Answer: Oscillating water columns function as wave energy converters and self-powered actuators in marine robotics. They harness wave motion to generate energy for autonomous operation in the ocean and remote monitoring.

132. How can robots optimize multi-material laminated actuators for shape-changing and adaptable structures in robotics and morphing aircraft?

Answer: Multi-material laminated actuators enable shape-changing and adaptable structures in robotics and morphing aircraft. Optimizing their design involves material selection and integration techniques for efficient actuation.

133. What are the advantages of using electroactive paper (Papp) actuators in flexible electronics and bio-inspired robots for delicate tasks?

Answer: Papp actuators offer advantages in flexible electronics and bio-inspired robots. They provide lightweight and responsive motion for delicate tasks like artificial insects and medical devices.

134. How do robots employ microelectromechanical systems (MEMS) deformable mirrors as adaptive optics for correcting optical aberrations in telescopes and laser systems?

Answer: MEMS deformable mirrors serve as adaptive optics in telescopes and laser systems. They adjust the shape of the mirror surface to correct optical aberrations, improving imaging and laser beam quality.

135. What is the significance of conducting polymer actuators in soft robotics for tactile sensing and responsive gripping in delicate tasks?

Answer: Conducting polymer actuators are crucial in soft robotics for tactile sensing and responsive gripping. They enable robots to interact delicately with objects and environments for various applications.

136. How can robots benefit from the use of micro hydraulic actuators for high-force and high-precision motion control in micro-robotics and medical devices?

Answer: Micro hydraulic actuators provide high-force and high-precision motion control in micro-robotics and medical devices. They enhance dexterity and strength for tasks like microsurgery and micro assembly.

137. What challenges do robots face when implementing liquid crystal elastomer (LCE) actuators for soft robotics and stretchable wearable devices?

Answer: Implementing LCE actuators in soft robotics and stretchable wearables involves challenges related to material compatibility and control for achieving stretchable and adaptable structures.

138. How do robots employ magnetostrictive thin-film actuators for precise fluid manipulation in microfluidic systems and lab-on-a-chip devices?

Answer: Magnetostrictive thin-film actuators enable precise fluid manipulation in microfluidic systems and lab-on-a-chip devices. They offer controlled and rapid adjustments for microscale applications.

139. What are the advantages of using Ferro fluidic rotary actuators in haptic feedback systems and virtual reality applications?

Answer: Ferro fluidic rotary actuators provide advantages in haptic feedback systems and virtual reality applications. They deliver immersive sensations and precise feedback for enhanced user experiences.

140. How can robots optimize shape memory alloy composites for versatile and adaptive structures in space exploration and robotics?

Answer: Shape memory alloy composites offer versatility and adaptability in space exploration and robotics. Optimizing their design involves integrating shape memory alloys with compatible materials for dynamic structures.

141. What role do thermoelectric actuators play in energy harvesting and self-powered sensors for remote monitoring and environmental sensing?

Answer: Thermoelectric actuators serve in energy harvesting and self-powered sensors for remote monitoring. They convert temperature gradients into electrical energy for autonomous data collection.

142. How do robots utilize fluidic elastomer actuators (FEAs) for soft robotics and compliant gripping in unstructured environments?

Answer: FEAs are essential in soft robotics for compliant gripping in unstructured environments. They enable safe and adaptable interactions with objects and terrain.

143. What is the significance of hydraulically amplified self-healing electrostatic (HASEL) actuators in soft robotics for versatile motion and manipulation?

Answer: HASEL actuators are significant in soft robotics for versatile motion and manipulation. They provide flexibility and robustness for tasks like soft grippers and adaptive structures.

144. How can robots benefit from the use of twisted and coiled actuators (TCAs) for bio-inspired locomotion and biomimetic robots in challenging terrains?

Answer: TCAs offer benefits in bio-inspired locomotion and biomimetic robots for traversing challenging terrains. They enable agile and adaptable movements in complex environments.

145. What challenges do robots face when implementing dielectric elastomer stack actuators for high-force and precise actuation in robotics and automation?

Answer: Implementing dielectric elastomer stack actuators for high-force and precise actuation involves challenges related to control strategies and material performance for achieving efficient motion.

146. How can robots optimize soft pneumatic bellows actuators for compliant and safe interaction with humans in collaborative settings?

Answer: Soft pneumatic bellows actuators can be optimized for compliant and safe human-robot collaboration by fine-tuning control parameters and developing effective safety mechanisms.

147. What are the considerations for using dielectric elastomer generators (DEGs) as self-powered actuators for vibration energy harvesting in industrial monitoring systems?

Answer: Implementing DEGs as self-powered actuators for vibration energy harvesting in industrial monitoring systems requires attention to device efficiency and integration for reliable operation.

148. How do robots employ bitableactuators for energy-efficient and stable position control in robotics and mechatronics?

Answer: Bitable actuators provide energy-efficient and stable position control in robotics and mechatronics. They maintain positions without constant power input, enhancing energy efficiency.

149. What is the role of hydrostatic actuators in heavy-duty construction equipment and robotic arms for high force and precision?

Answer: Hydrostatic actuators play a critical role in heavy-duty construction equipment and robotic arms. They offer high force and precision for tasks like lifting and excavation.

150. How can robots utilize smart fluids as actuators for adjustable damping and stiffness in automotive suspension systems and vibration control?

Answer: Smart fluids serve as actuators for adjustable damping and stiffness in automotive suspension systems. They enhance ride comfort and vehicle stability by adapting to road conditions.

151. What advantages do electrorheological (ER) actuators offer in precision positioning and force control in robotics and industrial automation?

Answer: ER actuators provide advantages in precision positioning and force control in robotics and industrial automation. They enable rapid adjustments for accurate and controlled movements.

152. How do robots employ magnetorheological (MR) actuators for adaptive structures and haptic feedback in virtual reality (VR) applications?

Answer: MR actuators enable adaptive structures and haptic feedback in VR applications. They create immersive sensations and responsive environments for enhanced VR experiences.

153. What are the challenges in controlling shape memory alloy fibers for shape-changing and deployable structures in aerospace and robotics?

Answer: Controlling shape memory alloy fibers for shape-changing and deployable structures involves addressing challenges related to heat management and actuation synchronization for reliable performance.

154. How can robots optimize pneumatic artificial muscles (PAMs) for assistive exoskeletons and rehabilitation devices?

Answer: Optimizing PAMs for assistive exoskeletons and rehabilitation devices involves customized design and control algorithms to enhance mobility and therapeutic support.

155. What is the significance of biohybrid actuators in bio-inspired robots and prosthetic devices for natural and lifelike movements?

Answer: Biohybrid actuators are significant in bio-inspired robots and prosthetic devices for achieving natural and lifelike movements. They mimic biological systems for enhanced functionality and user experience.

156. How do robots utilize twisted and coiled polymer actuators (TCPAs) for soft robotics and adaptable structures in unpredictable environments?

Answer: TCPAs are employed in soft robotics and adaptable structures for unpredictable environments. They enable robots to navigate and manipulate in dynamic and unstructured settings.

157. What role do helical hydraulic actuators play in advanced manufacturing processes and high-precision machining?

Answer: Helical hydraulic actuators are crucial in advanced manufacturing processes and high-precision machining. They provide controlled motion for intricate and detailed operations.

158. How can robots optimize conductive polymer actuators for responsive feedback and tactile sensing in human-robot interaction?

Answer: Optimizing conductive polymer actuators for responsive feedback and tactile sensing in human-robot interaction involves developing sensor integration and control strategies for enhanced communication and safety.

159. What are the advantages of using smart material actuators in deployable structures and space applications?

Answer: Smart material actuators offer advantages in deployable structures and space applications. They provide lightweight and efficient solutions for mission-critical deployments and adjustments.

160. How do robots employ dielectric elastomer stack actuators for adaptive optics and precise mirror control in astronomical telescopes?

Answer: Robots use dielectric elastomer stack actuators for adaptive optics and precise mirror control in astronomical telescopes. They correct atmospheric distortions for improved image quality and observations.

161. What is the role of magnetostrictive micro actuators in microfluidic systems for controlled fluid mixing and drug delivery applications?

Answer: Magnetostrictive micro actuators play a crucial role in microfluidic systems for controlled fluid mixing and drug delivery. They provide precise manipulation of fluids at the microscale, enabling efficient drug dispensing and chemical analysis.

162. How can robots benefit from the use of soft electroactive actuators for bio-inspired locomotion and adaptable structures in natural disaster response?

Answer: Soft electroactive actuators offer robots advantages in bio-inspired locomotion and adaptable structures for navigating disaster-stricken areas. Their compliance and adaptability enable efficient search and rescue operations.

163. What challenges do robots face when implementing shape memory alloy springs for actuation in deep-sea exploration and underwater robotics?

Answer: Implementing shape memory alloy springs in deep-sea exploration and underwater robotics presents challenges related to corrosion resistance and reliable actuation in extreme underwater environments.

164. How do robots employ phase-change actuation for reversible shape transformations in soft robotics and deployable structures?

Answer: Robots use phase-change actuation for reversible shape transformations in soft robotics and deployable structures. It enables shape-changing capabilities and adaptability for various applications.

165. What are the advantages of using hydraulically amplified piezoelectric actuators in precision manufacturing and micro-assembly tasks?

Answer: Hydraulically amplified piezoelectric actuators offer advantages in precision manufacturing and micro-assembly by combining high precision with significant force output, enhancing micro-scale manipulation.

166. How can robots optimize electroactive polymer actuators for responsive motion and tactile feedback in haptic interfaces and virtual reality (VR) devices?

Answer: Robots optimize electroactive polymer actuators for responsive motion and tactile feedback in haptic interfaces and VR devices through advanced control algorithms and material enhancements for immersive experiences.

167. What is the role of fluidic muscle actuators in rehabilitation robotics for assisting individuals in regaining mobility and strength?

Answer: Fluidic muscle actuators play a crucial role in rehabilitation robotics by providing assistance in regaining mobility and strength. Their compliance and adaptability aid in therapeutic exercises and support.

168. How do robots employ thermoelectric generators as self-powered actuators for remote sensing and monitoring in harsh environments?

Answer: Robots utilize thermoelectric generators as self-powered actuators for remote sensing and monitoring in harsh environments. They convert temperature differentials into electrical energy for autonomous operation.

169. What challenges do robots face when implementing hydrogel-based soft actuators for biohybrid robotics and biomedicine?

Answer: Implementing hydrogel-based soft actuators in biohybrid robotics and biomedicine involves challenges related to biocompatibility, control, and integration for safe and effective use in living organisms.

170. How can robots benefit from the use of adaptive airfoil actuators for aerodynamic optimization and efficiency in aerial vehicles and drones?

Answer: Robots benefit from adaptive airfoil actuators for aerodynamic optimization and efficiency in aerial vehicles and drones. They enable real-time adjustments for improved flight performance and energy efficiency.

171. What is the significance of biocompatible shape memory alloy actuators in medical robotics for minimally invasive surgeries and therapeutic interventions?

Answer: Biocompatible shape memory alloy actuators are significant in medical robotics for minimally invasive surgeries and therapeutic interventions. They offer precise control and compatibility with the human body for improved healthcare procedures.

172. How do robots utilize multilayered dielectric elastomer actuators for energy-efficient motion and adaptable structures in soft robotics and wearables?

Answer: Robots employ multilayered dielectric elastomer actuators for energy-efficient motion and adaptable structures in soft robotics and wearables. They enable flexible and responsive designs for various applications.

173. What are the challenges in controlling fiber-reinforced elastomer actuators for biomimetic propulsion and marine robots?

Answer: Controlling fiber-reinforced elastomer actuators for biomimetic propulsion and marine robots involves addressing challenges related to underwater operation and fluid dynamics for efficient locomotion.

174. How can robots optimize magnetic field-driven actuators for non-contact and precise manipulation in micro-robotics and lab-on-a-chip devices?

Answer: Robots optimize magnetic field-driven actuators for non-contact and precise manipulation in micro-robotics and lab-on-a-chip devices through advanced magnetic control systems and microfabrication techniques.

175. What role do smart fluids serve as actuators in adaptive optics and precise wave front correction in astronomical telescopes?

Answer: Smart fluids play a vital role as actuators in adaptive optics and precise wave front correction in astronomical telescopes. They correct atmospheric distortions for clearer and sharper observations.

176. How do robots employ thermally responsive polymer actuators for thermal regulation and temperature control in consumer electronics and medical devices?

Answer: Robots utilize thermally responsive polymer actuators for thermal regulation and temperature control in consumer electronics and medical devices. They maintain optimal operating conditions for improved performance and user comfort.

177. What are the advantages of using ionic polymer-metal composites (IPMCs) as actuators for microfluidic manipulation and lab-on-a-chip systems?

Answer: IPMCs offer advantages as actuators for microfluidic manipulation and lab-on-a-chip systems due to their small size, flexibility, and responsiveness, enhancing precision in fluid handling and analysis.

178. How can robots benefit from the use of conductive fabric actuators for flexible and wearable technology applications?

Answer: Robots benefit from conductive fabric actuators for flexible and wearable technology applications by enabling comfortable and responsive interfaces for health monitoring and communication.

179. What is the role of soft dielectric elastomer actuators (DEAs) in soft robotics for gentle object manipulation and safe human-robot interaction?

Answer: Soft DEAs play a vital role in soft robotics for gentle object manipulation and safe human-robot interaction. They provide compliance and adaptability for handling delicate objects and collaborating with humans.

180. How do robots employ magnetorheological (MR) fluid-based actuators for variable damping and vibration control in automotive suspension systems?

Answer: Robots employ MR fluid-based actuators for variable damping and vibration control in automotive suspension systems. They optimize ride comfort and vehicle stability by adjusting damping characteristics.

181.What challenges do robots face when implementing conductive polymer actuators for precise motion control in MEMS devices and micro-robotics?

Answer: Implementing conductive polymer actuators for precise motion control in MEMS devices and micro-robotics involves challenges related to miniaturization, control algorithms, and material performance.

182. How can robots optimize shape memory alloy wires for actuation in miniature surgical robots and medical implants?

Answer: Robots optimize shape memory alloy wires for actuation in miniature surgical robots and medical implants through advanced control systems and biocompatible materials for precise and safe procedures.

183. What are the advantages of using twisted and coiled actuators (TCAs) as artificial muscles in soft robotics for lifelike motion and dexterity?

Answer: TCAs offer advantages as artificial muscles in soft robotics for lifelike motion and dexterity. They enable robots to mimic natural movements and interact effectively with their environment.

184. How do robots employ micro hydraulic actuators for high-force and high-speed motion control in industrial automation and material handling?

Answer: Robots employ micro hydraulic actuators for high-force and high-speed motion control in industrial automation and material handling. They enhance efficiency and productivity in manufacturing processes.

185. What is the significance of soft pneumatic bellows actuators in wearable exoskeletons and assistive devices for enhancing mobility and strength?

Answer: Soft pneumatic bellows actuators are significant in wearable exoskeletons and assistive devices for enhancing mobility and strength. They provide support and assistance for individuals with mobility impairments.

186. How can robots optimize twisted and coiled polymer actuators (TCPAs) for adaptive structures and shape-changing in aerospace and deployable systems?

Answer: Robots optimize TCPAs for adaptive structures and shape-changing in aerospace and deployable systems by integrating them into lightweight and robust designs for efficient actuation.

187. What role do hydrostatic actuators serve in high-precision machining and heavy-duty industrial applications for accurate positioning and force control?

Answer: Hydrostatic actuators play a vital role in high-precision machining and heavy-duty industrial applications for accurate positioning and force control, ensuring precise and efficient operations.

188. How do robots utilize magnetorheological (MR) fluid-based actuators for haptic feedback and realistic tactile sensations in virtual reality (VR) environments?

Answer: Robots utilize MR fluid-based actuators for haptic feedback and realistic tactile sensations in VR environments. They enhance immersion and user engagement in virtual experiences.

189. What are the challenges in controlling bitable actuators for energy-efficient and stable position control in robotics and automation?

Answer: Controlling bitable actuators for energy-efficient and stable position control in robotics and automation involves addressing challenges related to feedback control and actuation synchronization.

190. How can robots benefit from the use of electrorheological (ER) actuators for precise force control and manipulation in robotic surgery and teleoperation?
Answer: Robots benefit from ER actuators for precise force control and manipulation in robotic surgery and teleoperation. They offer accurate and responsive movements for delicate procedures.

ONLINE RESOURCES

Practice Paper Link: https://forms.gle/8bfoKcWqWj6vtaoe9

QR Code:

Video Content Link: https://www.youtube.com/playlist?list=PLzlh_31VWEFoBWb-nAVnuBGmhYnCyAVhw

QR Code:

Chapter 7: Robot Vision and Perception

1. How do robots use cameras and sensors for visual perception?

Answer: Robots use cameras and sensors to capture visual data from their surroundings, allowing them to perceive and understand their environment.

2. What is the role of computer vision in robotics?

Answer: Computer vision in robotics processes visual data to identify objects, recognize patterns, and make informed decisions.

3. How do depth-sensing technologies like LiDAR assist in 3D perception?

Answer: LiDAR technology measures distance using laser pulses, providing robots with 3D perception and object mapping capabilities.

4. What are the advantages of using neural networks for image recognition in robotics?

Answer: Neural networks enable robots to recognize and classify objects in images, enhancing their ability to interact with the environment.

5. How does sensor fusion improve a robot's perception of its surroundings?

Answer: Sensor fusion combines data from multiple sensors, such as cameras and IMUs, to provide a more comprehensive and accurate perception of the environment.

6. What is the significance of SLAM (Simultaneous Localization and Mapping) in robotic navigation?

Answer: SLAM allows robots to create maps of unknown environments while simultaneously determining their own position, enabling autonomous navigation.

7. How do robots handle occlusions in visual perception?

Answer: Robots use algorithms to predict and compensate for occlusions, ensuring they can perceive objects even when partially hidden.

8. What role does object detection play in robotics?

Answer: Object detection algorithms enable robots to identify and locate specific objects within their visual field, facilitating tasks like object manipulation.

9. How do convolutional neural networks (CNNs) contribute to image analysis in robotics?

Answer: CNNs are specialized neural networks used for image analysis, enabling robots to extract features and make sense of visual data.

10. What challenges do robots face in recognizing and understanding human gestures?

Answer: Challenges in recognizing human gestures include variations in poses and lighting conditions, which robots must account for in their perception systems.

11. How can robots recognize and respond to facial expressions in human-robot interaction?

Answer: Robots employ facial recognition algorithms to detect and interpret human facial expressions, enabling more natural and responsive interactions.

12. What are the applications of 3D vision systems in industrial robotics?

Answer: 3D vision systems are used in industrial robotics for tasks such as quality control, bin picking, and assembly line automation.

13. How do robots adapt to changing lighting conditions for consistent perception?

Answer: Robots use adaptive algorithms and sensors to adjust to varying lighting conditions, maintaining accurate perception.

14. What is the significance of object tracking in robotic surveillance systems?

Answer: Object tracking allows robots to monitor and follow objects of interest, enhancing security and surveillance capabilities.

15. How do robots employ machine learning for scene understanding in autonomous vehicles?

Answer: Machine learning algorithms enable autonomous vehicles to understand complex traffic scenes, enhancing safety and navigation.

16. What are the challenges in achieving real-time image processing in robotics?

Answer: Real-time image processing in robotics requires high computational power and efficient algorithms to process visual data quickly.

17. How can robots recognize and interpret natural language commands for improved human-robot communication?

Answer: Natural language processing (NLP) techniques enable robots to understand and respond to spoken or written instructions from humans.

18. What role does image segmentation play in robotics?

Answer: Image segmentation divides an image into distinct regions, helping robots identify and separate objects in visual data.

19. How do robots use depth perception for obstacle avoidance in autonomous navigation?

Answer: Depth perception helps robots detect obstacles and plan safe paths, crucial for autonomous navigation and collision avoidance.

20. What are the ethical considerations in robot vision and perception, particularly in privacy-sensitive areas?

Answer: Ethical concerns in robot vision include privacy issues when capturing and processing visual data, necessitating responsible data handling and protection measures.

21. How can robots overcome challenges like sensor noise and calibration errors in perception?

Answer: Calibration and sensor fusion techniques are used to reduce noise and correct errors in perception systems, improving accuracy.

22. What are the benefits of using thermal imaging in robot vision for night vision and temperature detection?

Answer: Thermal imaging allows robots to operate in low-light conditions, detect temperature variations, and identify objects based on their heat signatures.

23. How do robots recognize and interpret road signs and traffic signals in autonomous vehicles?

Answer: Machine learning algorithms enable autonomous vehicles to recognize and interpret road signs and traffic signals, ensuring safe and compliant driving.

24. What is the role of object recognition in logistics and warehouse automation?

Answer: Object recognition systems help robots identify and categorize products in logistics and warehouse environments, streamlining inventory management and order fulfillment.

25. How can robots enhance agricultural practices through image analysis for crop monitoring and pest detection?

Answer: Robots use image analysis to monitor crops, detect pests, and optimize agricultural practices, leading to improved crop yields and sustainability.

26. How do robots address challenges like occlusions and cluttered scenes in visual perception for search and rescue operations?

Answer: Robots in search and rescue operations use advanced algorithms to handle occlusions and navigate cluttered environments, aiding in locating survivors.

27. What are the safety measures in place to prevent accidents in human-robot collaboration scenarios involving robot vision?

Answer: Safety measures include implementing collision detection systems, ensuring slow and cautious movements, and using protective barriers when robots and humans work together.

28. How do robots recognize and classify defects in manufacturing processes using visual inspection?

Answer: Robots employ visual inspection systems to identify and classify defects in manufacturing, improving product quality and reducing waste.

29. What are the considerations for designing accessible interfaces for visually impaired users in robot applications?

Answer: Considerations include providing auditory or tactile feedback and implementing voice commands to ensure visually impaired users can interact effectively with robots.

30. How can robots adapt to dynamic environments and changing objects in their visual field for robust perception?
Answer: Adaptive algorithms and continuous learning enable robots to adapt to dynamic environments and changing objects, ensuring robust perception.
31. How do robots use gesture recognition for intuitive control in human-robot interaction?
Answer: Gesture recognition technology allows robots to interpret human gestures as commands, enhancing natural and intuitive communication.
32. What challenges do robots face in recognizing and interpreting emotions in human facial expressions?
Answer: Challenges in emotion recognition include the complexity of human emotions and the need for robust algorithms to accurately interpret facial expressions.
33. How do robots employ scene reconstruction techniques for 3D modeling and virtual reality applications?
Answer: Scene reconstruction techniques help robots create 3D models of their environment, which can be used in virtual reality applications for immersive experiences.
34. What is the role of image stitching in panoramic photography and 360-degree vision systems?
Answer: Image stitching combines multiple images to create panoramic views or 360-degree vision, useful for surveillance, navigation, and entertainment.
35. How do robots use visual odometer for estimating their position and trajectory in GPS-denied environments?
Answer: Visual odometer relies on visual cues to estimate a robot's position and trajectory when GPS signals are unavailable, ensuring accurate navigation.
36. How do robots employ hyperspectral imaging for precise material identification in environmental monitoring and agriculture?
Answer: Robots use hyperspectral imaging to capture data across a wide range of wavelengths, enabling them to identify specific materials, detect pollutants, and assess crop health in environmental and agricultural applications.
37. What is the role of edge detection algorithms in robotics for contour recognition and object boundary delineation?
Answer: Edge detection algorithms highlight object edges in images, aiding in contour recognition and object boundary delineation, which are vital for tasks such as object manipulation and recognition.
38. How can robots utilize gaze tracking technology for human-robot interaction and enhancing user engagement?
Answer: Gaze tracking technology allows robots to monitor and respond to a user's eye movements and focus points, enhancing human-robot interaction and tailoring responses to user interests.
39. What is the significance of scene understanding in autonomous vehicles for safe and efficient driving in complex traffic environments?
Answer: Scene understanding in autonomous vehicles involves analyzing the road, traffic, and pedestrians. It's crucial for safe navigation, allowing vehicles to make informed decisions and avoid accidents.
40. How do robots apply feature extraction techniques in image processing to identify and characterize key elements in visual data?
Answer: Feature extraction techniques extract meaningful information from visual data, helping robots identify and characterize objects, patterns, or attributes for various applications, including object recognition and classification.
41. What role does object pose estimation play in robotic manipulation and pick-and-place tasks?
Answer: Object pose estimation determines the position and orientation of objects, enabling robots to grasp and manipulate them accurately in tasks like pick-and-place operations.
42. How can robots overcome challenges like image distortion and lens aberrations in visual perception?
Answer: Robots use calibration and image correction techniques to mitigate image distortion and lens aberrations, ensuring accurate perception in challenging visual conditions.
43. What is the significance of semantic segmentation in robotics for understanding scene context and object relationships?

Answer: Semantic segmentation classifies pixels in an image into meaningful categories, helping robots understand scene context and object relationships, crucial for navigation and interaction.

44. How do robots employ generative adversarial networks (GANs) for image synthesis and data augmentation in training perception models?

Answer: GANs generate synthetic images that enhance training data diversity for perception models, improving their robustness and ability to handle various scenarios.

45. What challenges do robots face in recognizing and navigating through unstructured natural environments using vision?

Answer: Challenges include handling diverse terrains, dynamic obstacles, and varying lighting conditions. Robots need advanced algorithms and sensors to navigate safely in such environments.

46. How can robots enhance safety in autonomous driving through advanced vision-based systems for pedestrian detection and collision avoidance?

Answer: Vision-based systems identify pedestrians and potential collision risks in real-time, allowing autonomous vehicles to take evasive actions and improve road safety.

47. What is the role of visual SLAM in augmented reality (AR) applications for overlaying virtual objects onto the real world?

Answer: Visual SLAM provides AR applications with real-time tracking of the user's environment, enabling the seamless integration of virtual objects into the real world.

48. How do robots adapt to diverse cultural gestures and expressions in human-robot interaction for global accessibility?

Answer: Robots utilize machine learning and cross-cultural training datasets to understand and respond to diverse cultural gestures and expressions, ensuring inclusive and effective communication.

49. What are the advantages of using hyperspectral imaging in precision agriculture for monitoring crop health and optimizing resource usage?

Answer: Hyperspectral imaging allows robots in agriculture to assess crop health based on spectral signatures, enabling precise resource allocation and improving crop yield and sustainability.

50. How can robots employ visual place recognition for localization and mapping in GPS-denied environments like indoor spaces?

Answer: Visual place recognition uses images to recognize and localize within previously visited environments, making it a valuable tool for robot navigation in indoor settings.

51. What challenges do robots face in recognizing and interpreting non-verbal cues like body language in human-robot interaction?

Answer: Challenges include the complexity and cultural variations of body language. Robots need robust algorithms to accurately interpret non-verbal cues in diverse social interactions.

52. How do robots use image compression techniques to optimize storage and transmission of visual data in remote sensing and surveillance applications?

Answer: Image compression reduces the size of visual data while preserving essential information, enabling efficient storage and transmission in remote sensing and surveillance.

53. What is the role of visual tracking in robotics for following moving objects or targets in real-time?

Answer: Visual tracking allows robots to monitor and follow moving objects, making it valuable in applications like drone tracking, surveillance, and sports analysis.

54. How can robots apply gesture recognition for intuitive control in smart home automation and human-robot collaboration?

Answer: Gesture recognition enables users to control robots and smart home devices through hand gestures, simplifying interaction and enhancing automation.

55. What are the privacy and security considerations in robot vision systems, especially in applications involving video surveillance?

Answer: Privacy and security measures include data encryption, access control, and anonymization to protect individuals' privacy and prevent unauthorized access to visual data in surveillance applications.

56. How do robots employ stereo vision for depth perception and 3D reconstruction in tasks like object manipulation and obstacle avoidance?

Answer: Stereo vision uses two cameras to capture depth information, allowing robots to perceive the 3D structure of their surroundings and make informed decisions.

57. What are the benefits of using event-based cameras in robotics for high-speed motion detection and low latency?

Answer: Event-based cameras offer high temporal resolution and low latency, making them suitable for tasks that require rapid motion detection and response, such as drone racing and collision avoidance.

58. How can robots recognize and interpret handwritten or printed text in documents and signage for information extraction?

Answer: Robots use optical character recognition (OCR) technology to recognize and extract text from documents and signage, enabling data processing and language understanding.

59. What is the significance of image registration in medical robotics for aligning pre-operative and intra-operative images during surgeries?

Answer: Image registration ensures the alignment of pre-operative and intra-operative images, assisting surgeons in precise navigation and interventions during medical procedures.

60. How do robots employ facial emotion recognition in applications like healthcare for patient interaction and emotional support?

Answer: Robots with facial emotion recognition capabilities can assess and respond to patients' emotional states, offering support and enhancing the healthcare experience.

61. How do robots apply visual scene understanding in disaster response scenarios to assess the severity of damage and plan rescue operations?

Answer: In disaster response, visual scene understanding enables robots to analyze the extent of damage, identify hazards, and plan rescue operations efficiently. By interpreting images and videos, robots can locate survivors, prioritize rescue efforts, and ensure the safety of both responders and victims.

62. What challenges do robots face in recognizing and understanding complex human actions and interactions for social robotics applications?

Answer: Recognizing and understanding complex human actions and interactions pose challenges in social robotics, including the need for sophisticated algorithms that account for contextual cues, gestures, and emotions. Robots must also adapt to diverse cultural norms and interpersonal variations to facilitate meaningful social interactions.

63. How can robots employ visual inspection techniques in manufacturing quality control for identifying defects and ensuring product consistency?

Answer: Visual inspection techniques play a crucial role in manufacturing quality control by enabling robots to detect defects, deviations, and anomalies in products. Robots use cameras and image analysis algorithms to ensure consistent quality and reduce production errors.

64. What is the role of facial recognition technology in enhancing security and access control in public spaces and secure facilities?

Answer: Facial recognition technology enhances security and access control by allowing robots and surveillance systems to identify individuals based on facial features. It is used in applications like airport security, access to secure facilities, and public safety.

65. How do robots utilize visual SLAM for creating maps of indoor environments and navigating autonomously in GPS-denied areas?

Answer: Visual SLAM is instrumental in robot navigation in GPS-denied areas such as indoor environments. Robots use cameras and visual cues to simultaneously create maps of their surroundings and determine their own position, enabling accurate and autonomous navigation.

66. What are the applications of vision-based gesture recognition in gaming and virtual reality for immersive and interactive experiences?

Answer: Vision-based gesture recognition is used in gaming and virtual reality to allow users to interact with virtual environments using hand and body gestures. This technology enhances immersion, gameplay, and user engagement in virtual experiences.

67. How can robots employ image captioning algorithms to provide visually impaired users with detailed descriptions of their surroundings?

Answer: Robots equipped with image captioning algorithms can describe scenes and objects to visually impaired users, enhancing their understanding of the environment and promoting independent navigation and interaction.

68. What challenges do robots face when recognizing and interpreting emotions in human facial expressions for empathetic responses in healthcare and therapy applications?

Answer: Recognizing and interpreting emotions in human facial expressions for empathetic responses in healthcare and therapy applications is challenging due to the subtle variations in emotional expressions. Robots must employ advanced facial recognition and emotion analysis algorithms to provide appropriate and empathetic support.

69. How do robots use visual tracking for monitoring and studying wildlife behavior in conservation efforts and ecological research?

Answer: Visual tracking allows robots to observe and document wildlife behavior for conservation efforts and ecological research. Robots equipped with cameras and tracking algorithms can capture data on animal movements, interactions, and habitat use, contributing to wildlife management and research.

70. What is the role of object recognition in robotics for assisting visually impaired individuals in identifying everyday objects and obstacles?

Answer: Object recognition in robotics assists visually impaired individuals by identifying and describing everyday objects and obstacles in their surroundings. This technology enables greater independence and safety for individuals with visual impairments.

71. How can robots apply image processing techniques for enhancing medical imaging, including image denoising and feature extraction for diagnosis and treatment planning?

Answer: Robots employ image processing techniques such as denoising and feature extraction to enhance the quality of medical images. This improves the accuracy of diagnosis, treatment planning, and medical research, leading to better patient outcomes.

72. What are the ethical considerations in the use of facial recognition and visual surveillance technologies for public safety and law enforcement?

Answer: Ethical considerations in facial recognition and visual surveillance technologies include issues related to privacy, surveillance abuse, and bias. It is crucial to establish guidelines and regulations to ensure responsible and ethical use of these technologies in public safety and law enforcement.

73. How do robots use visual odometer and mapping for exploring and mapping unknown environments, such as caves and uncharted terrain?

Answer: Visual odometer and mapping enable robots to explore and map unknown environments by tracking their own movements and creating maps using visual data. This technology is valuable in applications like cave exploration and search and rescue operations in challenging terrain.

74. What role does object tracking play in autonomous drone operations for monitoring and following moving subjects in aerial photography and surveillance?

Answer: Object tracking is essential in autonomous drone operations for monitoring and following moving subjects in applications like aerial photography and surveillance. Drones equipped with tracking algorithms can capture dynamic scenes and ensure subjects remain in focus, improving the quality of aerial imagery.

75. How can robots enhance accessibility for individuals with cognitive impairments by using visual cues and augmented reality interfaces for assistance in daily tasks?

Answer: Robots can enhance accessibility for individuals with cognitive impairments by providing visual cues and augmented reality interfaces that assist with daily tasks. These technologies offer step-by-step guidance and support, promoting independence and quality of life.

76. What is the significance of image forgery detection in digital forensics for identifying manipulated or altered images and videos?

Answer: Image forgery detection is critical in digital forensics for identifying manipulated or altered images and videos. It helps investigators determine the authenticity of digital evidence and assess the credibility of visual content in legal cases.

77. How do robots employ visual serving for precise and real-time control of robotic manipulators in tasks like surgical procedures and industrial automation?

Answer: Visual serving allows robots to control their manipulators with precision and in real-time by using visual feedback. This technology is applied in surgical procedures, industrial automation, and other tasks requiring accurate and dynamic control.

78. What are the benefits of using facial recognition technology in access control and user authentication for enhanced security and convenience?

Answer: Facial recognition technology enhances security and convenience by providing a seamless and contactless means of access control and user authentication. It is used in applications like smartphone unlocking, secure access to buildings, and identity verification.

79. How can robots use gaze analysis to assess human attention and engagement for adaptive user interfaces and personalized experiences?

Answer: Gaze analysis allows robots to assess human attention and engagement, enabling adaptive user interfaces and personalized experiences. Robots can adjust their behavior and content presentation based on the user's gaze patterns, improving communication and interaction.

80. What is the role of vision-based emotion recognition in human-robot collaboration for tasks that require emotional sensitivity and social intelligence?

Answer: Vision-based emotion recognition enables robots to perceive and respond to human emotions, enhancing their ability to collaborate effectively in tasks that require emotional sensitivity and social intelligence, such as therapy and customer service.

81. How do robots apply visual analysis techniques in sports analytics for tracking player movements and analyzing game dynamics?

Answer: Visual analysis techniques in sports analytics use cameras and tracking algorithms to monitor player movements and analyze game dynamics. This technology provides valuable insights for coaches, analysts, and sports enthusiasts, enhancing performance evaluation and strategy development.

82. What challenges do robots face when recognizing and interpreting visual metaphors and abstract concepts in art and cultural contexts?

Answer: Recognizing and interpreting visual metaphors and abstract concepts in art and cultural contexts is challenging due to the subjective nature of interpretation. Robots must employ advanced image analysis and contextual understanding to appreciate and communicate about artistic and cultural expressions.

83. How can robots use visual inspection systems for quality assurance and defect detection in manufacturing and production lines?

Answer: Visual inspection systems in manufacturing allow robots to perform quality assurance and defect detection tasks. By analyzing visual data, robots can identify and reject defective products, ensuring product quality and consistency.

84. What is the significance of object tracking in augmented reality applications for overlaying digital information on real-world objects and scenes?

Answer: Object tracking is crucial in augmented reality applications for precisely overlaying digital information on real-world objects and scenes. This technology enhances the user's augmented reality experience by ensuring accurate alignment and interaction with virtual elements.

85. How do robots employ facial recognition in healthcare settings for patient identification and medical records access, ensuring accuracy and security?

Answer: Robots use facial recognition technology in healthcare for accurate patient identification and secure access to medical records. This streamlines administrative processes, reduces errors, and enhances patient data security.

86. What are the privacy and ethical considerations in using visual surveillance and biometric technologies for monitoring public spaces and individuals?

Answer: Privacy and ethical considerations in visual surveillance and biometric technologies include concerns about privacy infringement, data security, and potential misuse. It is essential to establish transparent policies and safeguards to protect individuals' rights and ensure responsible use of these technologies.

87. How can robots apply visual analysis and object recognition for inventory management and stock tracking in retail and warehouses?

Answer: Robots use visual analysis and object recognition to perform inventory management and stock tracking in retail and warehouses. By identifying products and their locations, robots optimize inventory control and order fulfillment processes.

88. What role does visual inspection play in the assessment of infrastructure and critical assets for maintenance and safety purposes?

Answer: Visual inspection is essential for assessing the condition of infrastructure and critical assets. Robots equipped with cameras and inspection tools can identify defects, corrosion, or damage, helping ensure maintenance and safety compliance.

89. How do robots employ anomaly detection algorithms in visual surveillance for identifying unusual or suspicious behavior in security and public safety applications?

Answer: Robots use anomaly detection algorithms in visual surveillance to identify unusual or suspicious behavior patterns, enhancing security and public safety efforts. This technology alerts security personnel to potential threats or incidents.

90. What is the role of visual recognition in assistive technology for individuals with visual impairments, including object and text recognition?

Answer: Visual recognition technology assists individuals with visual impairments by recognizing and describing objects, text, and scenes. This empowers them to better understand and navigate their surroundings, promoting independence and accessibility.

91. How can robots apply visual SLAM techniques for mapping and exploring underwater environments in marine research and exploration?

Answer: Visual SLAM techniques enable robots to map and explore underwater environments by using cameras and visual data. This technology supports marine research, underwater archaeology, and exploration missions in challenging aquatic settings.

92. What challenges do robots face when recognizing and interpreting hand gestures and sign language for improved communication with individuals who are deaf or hard of hearing?

Answer: Recognizing and interpreting hand gestures and sign language for communication with individuals who are deaf or hard of hearing pose challenges due to variations in sign language and the need for robust gesture recognition algorithms. Robots must adapt to different signing styles and accurately convey messages to bridge communication gaps.

93. How do robots utilize visual analysis in environmental monitoring and remote sensing for assessing natural disasters, climate change, and ecosystem health?

Answer: Robots use visual analysis in environmental monitoring and remote sensing to assess natural disasters, track climate change, and monitor ecosystem health. Cameras and image analysis tools provide valuable data for environmental research and conservation efforts.

94. What is the significance of image-based localization and navigation in robotics for applications like autonomous drones and mobile robots in GPS-denied environments?

Answer: Image-based localization and navigation are crucial for robots operating in GPS-denied environments. By analyzing visual data and recognizing landmarks, robots can determine their position and navigate accurately, enhancing autonomy and reliability.

95. How can robots employ visual inspection and recognition in agriculture for crop disease detection, yield estimation, and precision farming?

Answer: Robots use visual inspection and recognition in agriculture to detect crop diseases, estimate yields, and implement precision farming practices. This technology promotes efficient resource utilization and sustainable agriculture.

96. What are the benefits of using visual analytics in data exploration and decision support across various industries, including finance and healthcare?

Answer: Visual analytics allows industries like finance and healthcare to explore complex datasets, discover insights, and make informed decisions. Visual representations of data facilitate understanding and enhance decision-making processes.

97. How do robots apply visual tracking and recognition for human-robot collaboration in manufacturing and assembly tasks, ensuring safety and efficiency?

Answer: Visual tracking and recognition enable safe and efficient human-robot collaboration in manufacturing and assembly tasks. Robots can identify and track human workers, adjust their movements accordingly, and collaborate on tasks while maintaining safety standards.

98. What challenges do robots face when interpreting and responding to visual cues in art and cultural heritage preservation, such as recognizing artistic styles and historical context?

Answer: Interpreting and responding to visual cues in art and cultural heritage preservation is challenging due to the subjectivity and context-dependent nature of art. Robots must employ advanced image analysis, art history knowledge, and cultural understanding to recognize artistic styles, historical context, and cultural significance.

99. How can robots employ image-based gesture recognition for intuitive control and communication in smart homes and IoT devices?

Answer: Robots use image-based gesture recognition for intuitive control and communication in smart homes and IoT devices. Users can interact with devices and appliances by making gestures, simplifying home automation and enhancing user experience.

100. What is the role of visual analytics in predictive maintenance for identifying equipment failures and optimizing maintenance schedules in industrial settings?

Answer: Visual analytics supports predictive maintenance by analyzing data from industrial equipment, identifying potential failures, and optimizing maintenance schedules. This proactive approach reduces downtime, extends equipment lifespan, and enhances operational efficiency across various industries.

101. How do robots employ visual perception in autonomous exploration of unknown environments, such as space or deep-sea exploration?

Answer: Robots use visual perception to autonomously explore unknown environments like space and deep-sea. Cameras and image processing help them navigate, identify interesting features, and make informed decisions in challenging and remote terrains.

102. What challenges do robots face in recognizing and interpreting visual metaphors and symbols in user interfaces for intuitive human-robot interaction?

Answer: Recognizing and interpreting visual metaphors and symbols in user interfaces poses challenges in human-robot interaction. Robots must understand the meaning and context of symbols, ensuring effective communication and user-friendly interactions.

103. How can robots apply visual analytics in healthcare for medical image analysis, disease diagnosis, and treatment planning?

Answer: Visual analytics in healthcare supports medical image analysis, disease diagnosis, and treatment planning. By visualizing complex medical data, it assists clinicians in making accurate diagnoses and planning personalized treatments.

104. What is the role of image recognition in robotics for sorting and categorizing objects in recycling and waste management systems?

Answer: Image recognition in robotics plays a vital role in sorting and categorizing objects in recycling and waste management systems. Robots use cameras to identify and separate recyclable materials, improving recycling efficiency and reducing waste.

105. How do robots employ visual tracking for monitoring wildlife populations and behavior in conservation research and ecological studies?

Answer: Robots use visual tracking to monitor wildlife populations and behavior in conservation research and ecological studies. Cameras and tracking algorithms help gather data on animal movements, migratory patterns, and habitat use, contributing to wildlife conservation efforts.

106. What are the applications of facial recognition technology in the entertainment industry for personalized content recommendations and immersive experiences?

Answer: Facial recognition technology enhances personalized content recommendations and immersive experiences in the entertainment industry. It can identify viewers' emotions and preferences to recommend tailored content and create interactive and engaging experiences.

107. How can robots use visual analysis and object detection for safety and hazard detection in industrial environments and construction sites?

Answer: Robots employ visual analysis and object detection to enhance safety and hazard detection in industrial environments and construction sites. Cameras and algorithms identify potential safety risks and alert workers to take necessary precautions, reducing accidents and improving workplace safety.

108. What is the significance of image segmentation in medical robotics for precise tissue and organ delineation during surgeries and medical interventions?

Answer: Image segmentation in medical robotics is crucial for precise tissue and organ delineation during surgeries and interventions. It assists surgeons in accurately navigating and targeting specific areas, enhancing surgical precision and patient outcomes.

109. How do robots use visual perception in precision agriculture for monitoring crop growth, nutrient levels, and irrigation needs?

Answer: Robots employ visual perception in precision agriculture to monitor crop growth, assess nutrient levels, and determine irrigation needs. Cameras and image analysis aid in optimizing resource allocation and promoting sustainable farming practices.

110. What challenges do robots face in recognizing and responding to human emotional expressions in social robotics applications?

Answer: Recognizing and responding to human emotional expressions in social robotics applications pose challenges due to the complexity of emotions and individual variations. Robots need advanced emotional recognition algorithms and context awareness to provide appropriate responses and foster meaningful interactions.

111. How can robots apply image-based document analysis and recognition for digitizing and organizing paper documents in office and archival settings?

Answer: Robots use image-based document analysis and recognition to digitize and organize paper documents in office and archival settings. Optical character recognition (OCR) technology extracts text, making documents searchable and facilitating efficient data management.

112. What is the role of visual tracking in autonomous vehicles for monitoring other road users and ensuring safe navigation in traffic?

Answer: Visual tracking is essential in autonomous vehicles for monitoring other road users, including vehicles, pedestrians, and cyclists. It enables safe navigation by predicting and responding to the movements of surrounding objects and ensuring collision avoidance.

113. How do robots employ visual analysis in retail and e-commerce for product recognition, inventory management, and personalized shopping experiences?

Answer: Robots use visual analysis in retail and e-commerce to recognize products, manage inventory, and enhance personalized shopping experiences. Visual data helps identify products, track stock levels, and recommend items tailored to individual preferences.

114. What are the privacy and ethical considerations in using facial recognition and visual surveillance technologies in public spaces, workplaces, and consumer devices?

Answer: Privacy and ethical considerations in facial recognition and visual surveillance include concerns about data privacy, surveillance abuse, and bias. Implementing transparent policies, informed consent, and data protection measures is crucial to address these ethical concerns responsibly.

115. How can robots employ visual inspection and quality control in the automotive industry for identifying manufacturing defects and ensuring vehicle safety?

Answer: Robots use visual inspection and quality control in the automotive industry to identify manufacturing defects and ensure vehicle safety. Cameras and image analysis detect flaws, ensuring the production of safe and reliable vehicles.

116. What is the significance of image-based localization and mapping in robotics for applications like autonomous drones and exploration of remote environments?

Answer: Image-based localization and mapping are vital in robotics for applications like autonomous drones and exploration of remote environments. They enable robots to create maps and determine their positions by analyzing visual data, supporting navigation in challenging and unfamiliar terrains.

117. How do robots employ visual recognition in archaeological and cultural heritage preservation for artifact identification and restoration efforts?

Answer: Robots use visual recognition in archaeological and cultural heritage preservation to identify artifacts and assist in restoration efforts. Visual analysis helps archaeologists and conservators catalog, restore, and preserve historical and cultural treasures.

118. What role does image processing play in monitoring and controlling energy consumption in smart grid systems and sustainable infrastructure?

Answer: Image processing is essential in monitoring and controlling energy consumption in smart grid systems and sustainable infrastructure. It aids in analyzing visual data from sensors to optimize energy distribution and reduce environmental impact.

119. How can robots apply visual analytics in finance for fraud detection, risk assessment, and market analysis?

Answer: Robots employ visual analytics in finance for fraud detection, risk assessment, and market analysis. This technology helps identify irregularities, assess investment risks, and provide insights for financial decision-making.

120. What are the benefits of using visual inspection and pattern recognition in pharmaceutical manufacturing for quality assurance and compliance with regulatory standards?

Answer: Visual inspection and pattern recognition in pharmaceutical manufacturing enhance quality assurance and compliance with regulatory standards. Robots can identify defects, verify product consistency, and ensure pharmaceutical products meet safety and quality requirements.

121. How do robots use visual perception in underwater robotics for exploring and studying marine ecosystems and geological formations?

Answer: Robots utilize visual perception in underwater robotics to explore and study marine ecosystems and geological formations. Cameras and image analysis enable data collection and environmental monitoring, contributing to marine research and conservation.

122. What challenges do robots face when recognizing and interpreting non-verbal cues like micro expressions in human-robot interaction?

Answer: Recognizing and interpreting non-verbal cues like micro expressions in human-robot interaction pose challenges due to their subtle nature and individual variations. Robots must employ advanced emotion recognition algorithms and sensitivity to context to understand and respond appropriately to these cues.

123. How can robots employ visual analysis in logistics and supply chain management for tracking shipments, optimizing routes, and ensuring timely deliveries?

Answer: Robots use visual analysis in logistics and supply chain management to track shipments, optimize transportation routes, and ensure timely deliveries. Visual data aids in monitoring cargo and streamlining the logistics process.

124. What is the role of image recognition in robotics for autonomous navigation and obstacle avoidance in dynamic environments?

Answer: Image recognition in robotics plays a crucial role in autonomous navigation and obstacle avoidance in dynamic environments. It allows robots to identify and respond to obstacles in real-time, ensuring safe and efficient movement.

125. How do robots utilize visual perception for assisting individuals with visual impairments in reading text, recognizing faces, and navigating urban environments?

Answer: Robots employ visual perception to assist individuals with visual impairments by reading text, recognizing faces, and providing navigation support in urban environments. These capabilities enhance accessibility and independence for individuals with visual challenges.

126. What is the significance of image forgery detection in journalism and media for verifying the authenticity of visual content and news reports?

Answer: Image forgery detection is crucial in journalism and media for verifying the authenticity of visual content and news reports. It helps maintain journalistic integrity and ensures the accuracy of information presented to the public.

127. How can robots apply visual tracking and recognition for human-robot collaboration in healthcare, such as assisting with surgeries and patient care?

Answer: Robots use visual tracking and recognition for human-robot collaboration in healthcare, assisting with surgeries and patient care. These technologies enable robots to work alongside medical professionals, enhancing precision and patient outcomes.

128. What challenges do robots face when recognizing and interpreting symbols and icons in user interfaces for universal accessibility and usability?

Answer: Recognizing and interpreting symbols and icons in user interfaces for universal accessibility and usability pose challenges due to the need for cross-cultural understanding and context-awareness. Robots must adapt to different cultural interpretations and user preferences to provide inclusive and user-friendly experiences.

129. How can robots employ image-based gesture recognition for natural and intuitive human-robot interaction in education and training?

Answer: Robots utilize image-based gesture recognition for natural and intuitive human-robot interaction in education and training. This technology allows users to convey information and engage with robots through gestures, facilitating effective teaching and learning experiences.

130. What is the role of visual analytics in geospatial applications for analyzing satellite imagery, mapping terrain, and monitoring environmental changes?

Answer: Visual analytics plays a critical role in geospatial applications for analyzing satellite imagery, mapping terrain, and monitoring environmental changes. It provides insights for environmental research, disaster management, and land-use planning.

131. How do robots utilize visual recognition in smart cities for traffic management, pedestrian safety, and urban planning?

Answer: In smart cities, robots employ visual recognition to manage traffic flow, enhance pedestrian safety, and support urban planning. Cameras and image analysis help monitor road conditions, detect congestion, and analyze pedestrian movement, contributing to efficient city operations.

132. What challenges do robots face in interpreting and responding to visual art and aesthetics, including assessing artistic quality and creativity?

Answer: Interpreting and responding to visual art and aesthetics present challenges for robots due to the subjective nature of artistic evaluation. Robots need to combine image analysis with an understanding of art history and cultural context to assess artistic quality and creativity accurately.

133. How can robots apply visual tracking in sports coaching and analysis for capturing player performance and strategic insights?

Answer: Robots use visual tracking in sports coaching and analysis to capture player performance and strategic insights. Cameras and tracking algorithms help analyze player movements, contributing to game analysis and performance improvement.

134. What is the significance of image-based authentication and verification in biometric security systems for user access and identity protection?

Answer: Image-based authentication and verification are crucial in biometric security systems for user access and identity protection. By using facial recognition and other visual biometrics, these systems enhance security and protect against unauthorized access.

135. How do robots employ visual perception in the hospitality industry for guest services and personalized experiences?

Answer: In the hospitality industry, robots utilize visual perception for guest services and personalized experiences. They can recognize guests, provide information, and offer customized services, enhancing the overall guest experience.

136. What challenges do robots face when interpreting visual scenes and context in natural language processing for human-robot communication?

Answer: Interpreting visual scenes and context in natural language processing poses challenges for robots, as it requires integrating visual information with language understanding. Robots must develop the capability to understand and respond appropriately to visual cues described in human language.

137. How can robots use visual analysis in environmental conservation for monitoring wildlife populations, habitat changes, and ecological trends?

Answer: Robots apply visual analysis in environmental conservation to monitor wildlife populations, track habitat changes, and analyze ecological trends. Visual data aids in assessing the impact of environmental factors on ecosystems and guiding conservation efforts.

138. What is the role of image recognition in augmented reality for overlaying digital information on real-world objects and scenes?

Answer: Image recognition plays a vital role in augmented reality for accurately overlaying digital information on real-world objects and scenes. This technology ensures that virtual elements align seamlessly with the physical environment, enhancing the augmented reality experience.

139. How do robots employ visual tracking for autonomous drone navigation in aerial photography, mapping, and surveillance missions?

Answer: Visual tracking enables autonomous drones to navigate during aerial photography, mapping, and surveillance missions. Drones use cameras and tracking algorithms to follow subjects of interest, capture data, and ensure accurate imagery.

140. What are the benefits of using visual analytics in retail for customer behavior analysis, inventory optimization, and store layout planning?

Answer: Visual analytics offers benefits in retail by analyzing customer behavior, optimizing inventory, and planning store layouts. This technology helps retailers make data-driven decisions, enhance customer experiences, and improve operational efficiency.

141. How can robots apply image-based gesture recognition for enhancing accessibility in public spaces and facilitating communication with individuals with disabilities?

Answer: Robots use image-based gesture recognition to enhance accessibility in public spaces and communicate with individuals with disabilities. This technology enables non-verbal communication and interaction, making public spaces more inclusive.

142. What is the significance of image processing in autonomous vehicles for traffic sign recognition, lane detection, and obstacle avoidance?

Answer: Image processing is vital in autonomous vehicles for tasks such as traffic sign recognition, lane detection, and obstacle avoidance. Cameras and image analysis help autonomous vehicles interpret road conditions and navigate safely.

143. How do robots employ visual perception in the agricultural sector for crop monitoring, pest detection, and yield prediction?

Answer: Robots utilize visual perception in agriculture for crop monitoring, pest detection, and yield prediction. Cameras and image analysis aid in identifying crop health issues, optimizing pest control, and estimating crop yields, leading to more efficient farming practices.

144. What role does image recognition play in virtual fashion try-on applications for allowing users to visualize clothing and accessories?

Answer: Image recognition is essential in virtual fashion try-on applications, allowing users to visualize clothing and accessories. Users can virtually try on items and see how they look, enhancing the online shopping experience.

145. How can robots apply visual inspection and pattern recognition in pharmaceutical quality control for identifying medication packaging defects?

Answer: Robots use visual inspection and pattern recognition in pharmaceutical quality control to identify medication packaging defects. This technology ensures the safety and integrity of pharmaceutical products.

146. What is the role of image forgery detection in digital marketing and advertising for verifying the authenticity of visual content and campaigns?

Answer: Image forgery detection plays a role in digital marketing and advertising by verifying the authenticity of visual content and campaigns. It helps maintain trust and credibility in advertising efforts.

147. How do robots employ visual tracking and recognition in autonomous delivery systems for package identification and recipient interaction?

Answer: Robots use visual tracking and recognition in autonomous delivery systems for package identification and recipient interaction. Cameras and recognition algorithms ensure accurate deliveries and facilitate interactions with recipients.

148. What challenges do robots face when recognizing and interpreting abstract visual patterns and symbols for art and design applications?

Answer: Recognizing and interpreting abstract visual patterns and symbols pose challenges for robots in art and design applications due to the subjective nature of interpretation. Robots must develop an understanding of design principles and context to accurately interpret and create visual art.

149. How can robots apply visual analytics in urban planning for analyzing traffic patterns, optimizing transportation systems, and improving city infrastructure?

Answer: Robots use visual analytics in urban planning to analyze traffic patterns, optimize transportation systems, and improve city infrastructure. This technology supports data-driven decisions in urban development and management.

150. What is the significance of image-based localization and navigation in aerial drones for applications like agriculture, surveying, and environmental monitoring?

Answer: Image-based localization and navigation are significant in aerial drones for applications such as agriculture, surveying, and environmental monitoring. Drones rely on visual cues to accurately navigate and capture data, making them valuable tools in various industries.

151. How do robots utilize visual recognition in the education sector for enhancing interactive learning experiences and providing educational support?

Answer: In education, robots use visual recognition to enhance interactive learning experiences and provide educational support. They can identify objects, recognize student gestures, and adapt their teaching methods, fostering engagement and personalized learning.

152. What challenges do robots face in interpreting and responding to visual cues and body language in human-robot collaboration for teamwork and cooperation?

Answer: Interpreting and responding to visual cues and body language in human-robot collaboration pose challenges due to the nuances and cultural variations in non-verbal communication. Robots must develop sensitivity to these cues to promote effective teamwork and cooperation.

153. How can robots apply visual analysis in forensic science for evidence collection, crime scene reconstruction, and suspect identification?

Answer: Robots employ visual analysis in forensic science for evidence collection, crime scene reconstruction, and suspect identification. Cameras and image analysis tools aid investigators in solving crimes and ensuring accurate legal proceedings.

154. What is the role of image recognition in autonomous vehicles for pedestrian detection and enhancing road safety?

Answer: Image recognition plays a crucial role in autonomous vehicles by enabling pedestrian detection and enhancing road safety. Cameras and recognition algorithms help vehicles identify pedestrians and respond to potential hazards, reducing accidents.

155. How do robots employ visual tracking for precision agriculture in tasks such as plant counting, crop spraying, and soil analysis?

Answer: Robots utilize visual tracking in precision agriculture for tasks like plant counting, crop spraying, and soil analysis. Cameras and tracking algorithms assist in monitoring and managing crops, optimizing resource use and increasing agricultural efficiency.

156. What are the benefits of using visual analytics in healthcare for medical imaging interpretation, patient diagnosis, and treatment planning?

Answer: Visual analytics in healthcare offers benefits by aiding in medical imaging interpretation, patient diagnosis, and treatment planning. It enables healthcare professionals to make informed decisions, improving patient outcomes and care quality.

157. How can robots use image-based gesture recognition for enhancing human-robot interaction in customer service and hospitality?

Answer: Robots apply image-based gesture recognition to enhance human-robot interaction in customer service and hospitality. Recognizing and responding to gestures allows robots to assist guests and provide a more personalized and interactive experience.

158. What is the significance of image processing in satellite imagery analysis for environmental monitoring, disaster management, and climate research?

Answer: Image processing is significant in satellite imagery analysis for environmental monitoring, disaster management, and climate research. It helps extract valuable insights from satellite data, aiding in disaster response and environmental conservation.

159. How do robots employ visual perception in warehouse automation for tasks like inventory management, order picking, and logistics optimization?

Answer: Robots use visual perception in warehouse automation for inventory management, order picking, and logistics optimization. Cameras and image analysis assist in identifying and organizing products, streamlining warehouse operations.

160. What role does image recognition play in virtual reality for creating immersive and interactive environments?

Answer: Image recognition is essential in virtual reality for creating immersive and interactive environments. It allows virtual elements to respond to and interact with the physical world, enhancing the user's VR experience.

161. How can robots apply visual analysis in wildlife conservation for tracking endangered species, monitoring poaching activities, and preserving biodiversity?

Answer: Robots apply visual analysis in wildlife conservation to track endangered species, monitor poaching activities, and preserve biodiversity. Cameras and image processing support conservation efforts by collecting data and deterring illegal activities.

162. What is the role of image forgery detection in legal proceedings and digital evidence authentication?

Answer: Image forgery detection plays a crucial role in legal proceedings and digital evidence authentication. It ensures the integrity and credibility of visual evidence presented in court, helping establish the truth.

163. How do robots employ visual tracking and recognition in the entertainment industry for character animation and interactive experiences?

Answer: Robots utilize visual tracking and recognition in the entertainment industry for character animation and interactive experiences. Cameras and recognition technology enable characters to respond to users' actions and create immersive entertainment.

164. What challenges do robots face when interpreting and responding to facial expressions and emotional cues in healthcare for patient care and therapy?

Answer: Interpreting and responding to facial expressions and emotional cues in healthcare pose challenges for robots due to the complexity and variability of human emotions. Robots must develop emotional intelligence and adapt their responses to provide effective patient care and therapy.

165. How can robots apply visual analytics in marketing and advertising for consumer behavior analysis, campaign optimization, and content personalization?

Answer: Robots use visual analytics in marketing and advertising for consumer behavior analysis, campaign optimization, and content personalization. It helps marketers make data-driven decisions and deliver targeted content to consumers.

166. What is the significance of image-based authentication and biometric security in mobile devices and digital payment systems?

Answer: Image-based authentication and biometric security play a significant role in mobile devices and digital payment systems. They enhance user security and convenience by using facial recognition or fingerprint scanning for device access and payment authorization.

167. How do robots employ visual perception in underwater exploration for marine biology research and the study of underwater ecosystems?

Answer: Robots employ visual perception in underwater exploration for marine biology research and the study of underwater ecosystems. Cameras and image analysis contribute to data collection and environmental monitoring in challenging aquatic environments.

168. What challenges do robots face when recognizing and interpreting visual art styles and cultural symbolism in art history and museum curation?

Answer: Recognizing and interpreting visual art styles and cultural symbolism in art history and museum curation are challenging for robots due to the subjectivity and context-dependent nature of art. Robots must combine image analysis with cultural knowledge to accurately curate and appreciate art.

169. How can robots apply image recognition in remote sensing for agriculture to assess crop health, optimize irrigation, and predict crop yields?

Answer: Robots use image recognition in remote sensing for agriculture to assess crop health, optimize irrigation, and predict crop yields. Visual data aids in precision farming practices and resource management.

170. What is the role of image processing in autonomous drones for mapping and monitoring large-scale natural disasters and emergency response?

Answer: Image processing plays a vital role in autonomous drones for mapping and monitoring large-scale natural disasters and emergency response. It supports disaster management by providing real-time information and situational awareness to responders.

171. How do robots employ visual tracking for industrial automation in manufacturing processes, quality control, and material handling?

Answer: Robots use visual tracking for industrial automation in manufacturing processes, quality control, and material handling. Cameras and tracking technology help optimize production, maintain product quality, and streamline material flow.

172. What are the benefits of using visual analysis in architectural design for visualizing building concepts, assessing environmental impact, and optimizing construction processes?

Answer: Visual analysis in architectural design offers benefits by visualizing building concepts, assessing environmental impact, and optimizing construction processes. It enhances collaboration and decision-making in architecture and construction projects.

173. How can robots utilize visual perception in autonomous exploration of extreme environments, such as outer space or deep-sea exploration?

Answer: Robots utilize visual perception in autonomous exploration of extreme environments, including outer space and deep-sea exploration. Cameras and visual data analysis support navigation and data collection in challenging and remote terrains.

174. What is the significance of image forgery detection in journalism and media for preventing the spread of misinformation and fake news?

Answer: Image forgery detection is significant in journalism and media for preventing the spread of misinformation and fake news. It ensures the credibility of visual content and supports accurate reporting.

175. How do robots employ visual tracking and recognition in manufacturing for process optimization, error detection, and adaptive manufacturing?

Answer: Robots employ visual tracking and recognition in manufacturing for process optimization, error detection, and adaptive manufacturing. Cameras and recognition technology enable real-time monitoring and adjustment of manufacturing processes.

176. What challenges do robots face when interpreting and responding to visual cues in human-robot interaction for assisting individuals with autism or social communication difficulties?

Answer: Interpreting and responding to visual cues in human-robot interaction for assisting individuals with autism or social communication difficulties pose challenges. Robots must develop sensitivity to non-verbal cues and adapt their behavior to provide effective support and companionship.

177. How can robots apply image-based gesture recognition in automotive interfaces for driver assistance, infotainment control, and safety features?

Answer: Robots use image-based gesture recognition in automotive interfaces for driver assistance, infotainment control, and safety features. Recognizing driver gestures enhances the driving experience and safety by reducing distractions and facilitating hands-free control.

178. What is the role of visual analytics in aviation for analyzing flight data, pilot training, and aircraft maintenance?

Answer: Visual analytics plays a role in aviation by analyzing flight data, supporting pilot training, and optimizing aircraft maintenance. It enhances safety, efficiency, and decision-making in the aviation industry.

179. How do robots employ visual perception in archaeological excavations and artifact analysis for uncovering historical treasures and preserving cultural heritage?

Answer: Robots utilize visual perception in archaeological excavations and artifact analysis to uncover historical treasures and preserve cultural heritage. Visual data aids in documenting and studying artifacts, contributing to our understanding of the past.

180. What are the applications of facial recognition technology in security and access control for securing facilities and identifying potential threats?

Answer: Facial recognition technology has applications in security and access control for securing facilities and identifying potential threats. It enhances security by allowing authorized individuals access while detecting unauthorized or suspicious individuals.

181. How can robots apply visual analysis in the fashion industry for trend prediction, inventory management, and personalized shopping experiences?

Answer: Robots use visual analysis in the fashion industry to predict trends, manage inventory, and offer personalized shopping experiences. Visual data helps identify popular styles, optimize stock levels, and recommend fashion items tailored to individual preferences.

182. What challenges do robots face when interpreting and responding to visual and emotional cues in therapy and mental health support applications?

Answer: Interpreting and responding to visual and emotional cues in therapy and mental health support applications pose challenges for robots. Robots must develop empathy and emotional intelligence to provide effective support and companionship for individuals in need.

183. How can robots employ image recognition in agriculture for early pest detection, crop disease identification, and targeted treatment?

Answer: Robots apply image recognition in agriculture for early pest detection, crop disease identification, and targeted treatment. Visual data assists in monitoring crop health, enabling proactive pest control and minimizing chemical usage.

184. What is the role of image processing in autonomous navigation systems for drones and robots in search and rescue missions and disaster relief operations?

Answer: Image processing plays a vital role in autonomous navigation systems for drones and robots in search and rescue missions and disaster relief operations. It supports the rapid assessment of disaster-affected areas and assists in locating survivors.

185. How do robots employ visual tracking for sports analysis and broadcasting, including player performance evaluation and live event coverage?

Answer: Robots use visual tracking for sports analysis and broadcasting, including player performance evaluation and live event coverage. Cameras and tracking algorithms capture dynamic sports action, enhancing the viewing experience and providing valuable insights for coaches and fans.

186. What are the benefits of using visual analytics in finance for fraud prevention, investment portfolio management, and market forecasting?

Answer: Visual analytics in finance offers benefits by preventing fraud, optimizing investment portfolios, and forecasting market trends. It helps financial institutions make data-driven decisions, protect assets, and maximize returns for clients.

187. How can robots apply visual perception in the manufacturing industry for quality inspection, defect detection, and production process optimization?

Answer: Robots utilize visual perception in the manufacturing industry for quality inspection, defect detection, and production process optimization. Cameras and image analysis ensure product quality and streamline manufacturing processes.

188. What role does image recognition play in augmented reality applications for education, training simulations, and interactive learning experiences?

Answer: Image recognition is essential in augmented reality applications for education, training simulations, and interactive learning experiences. It allows users to interact with virtual content and enhances educational engagement.

189. How do robots employ visual analysis in wildlife monitoring and conservation for tracking animal populations, studying behavior, and protecting endangered species?

Answer: Robots employ visual analysis in wildlife monitoring and conservation for tracking animal populations, studying behavior, and protecting endangered species. Visual data contributes to scientific research and conservation efforts.

190. What is the significance of image forgery detection in law enforcement and criminal investigations for authenticating evidence and solving crimes?

Answer: Image forgery detection is significant in law enforcement and criminal investigations for authenticating evidence and solving crimes. It ensures the integrity of visual evidence, aiding in the pursuit of justice.

191. How can robots use visual tracking and recognition in the food industry for quality control, food safety inspection, and efficient packaging?

Answer: Robots apply visual tracking and recognition in the food industry for quality control, food safety inspection, and efficient packaging. Cameras and recognition technology help maintain food quality standards and improve production processes.

192. What challenges do robots face when recognizing and interpreting visual and non-verbal cues in human-robot collaboration for healthcare and rehabilitation?

Answer: Recognizing and interpreting visual and non-verbal cues in human-robot collaboration for healthcare and rehabilitation pose challenges. Robots must develop sensitivity to patients' emotional and physical cues to provide effective care and therapy.

193. How can robots employ image-based gesture recognition in smart home systems for intuitive control of appliances, lighting, and security?

Answer: Robots use image-based gesture recognition in smart home systems for intuitive control of appliances, lighting, and security. Recognizing gestures enhances home automation and improves user convenience.

194. What is the role of visual analytics in urban development for analyzing land use, traffic patterns, and infrastructure planning?

Answer: Visual analytics plays a role in urban development for analyzing land use, traffic patterns, and infrastructure planning. It supports informed decisions in city planning and development.

195. How do robots utilize visual perception in underwater archaeology for exploring submerged historical sites and uncovering maritime artifacts?

Answer: Robots utilize visual perception in underwater archaeology for exploring submerged historical sites and uncovering maritime artifacts. Cameras and visual data analysis aid in documenting and studying underwater cultural heritage.

196. What challenges do robots face when interpreting and responding to visual and auditory cues in human-robot interaction for individuals with disabilities?

Answer: Interpreting and responding to visual and auditory cues in human-robot interaction for individuals with disabilities pose challenges. Robots must integrate sensory information and adapt their communication methods to meet the diverse needs of users.

197. How can robots apply visual analysis in cultural heritage preservation for restoring artworks, artifacts, and historical buildings?

Answer: Robots use visual analysis in cultural heritage preservation for restoring artworks, artifacts, and historical buildings. Visual data aids conservators in assessing and preserving cultural treasures.

198. What is the significance of image-based authentication in cybersecurity for securing digital devices, accounts, and sensitive information?

Answer: Image-based authentication is significant in cybersecurity for securing digital devices, accounts, and sensitive information. It adds an extra layer of security by requiring visual recognition or biometric verification to access protected resources.

199. How do robots employ visual tracking and recognition in the aviation industry for aircraft maintenance, airport operations, and air traffic control?

Answer: Robots use visual tracking and recognition in the aviation industry for aircraft maintenance, airport operations, and air traffic control. Visual data assists in inspections, logistics, and ensuring safe and efficient air travel.

200. What are the benefits of using visual analytics in sports performance analysis for coaching, player development, and game strategy refinement?

Answer: Visual analytics in sports performance analysis offers benefits for coaching, player development, and game strategy refinement. It provides coaches and athletes with valuable insights, contributing to improved performance and strategy.

201. How can robots utilize image recognition in wildlife photography and conservation for identifying species, monitoring habitats, and raising awareness?

Answer: Robots apply image recognition in wildlife photography and conservation for identifying species, monitoring habitats, and raising awareness. Visual data supports conservation efforts and educates the public about biodiversity.

202. What role does image processing play in autonomous navigation for aerial drones in surveillance and reconnaissance missions?

Answer: Image processing plays a crucial role in autonomous navigation for aerial drones in surveillance and reconnaissance missions. It aids in capturing and analyzing visual data, enhancing situational awareness and mission effectiveness.

203. How do robots employ visual tracking for inventory management and logistics optimization in e-commerce warehouses?

Answer: Robots use visual tracking for inventory management and logistics optimization in e-commerce warehouses. Cameras and tracking technology help efficiently locate, pick, and pack products for shipment.

204. What is the role of image forgery detection in digital art authentication and the protection of intellectual property rights?

Answer: Image forgery detection plays a role in digital art authentication and the protection of intellectual property rights. It ensures the authenticity of digital artworks and helps prevent unauthorized use or reproduction.

205. How can robots apply visual perception in architecture and construction for site analysis, project visualization, and quality control?

Answer: Robots utilize visual perception in architecture and construction for site analysis, project visualization, and quality control. Visual data aids architects and builders in planning, monitoring progress, and ensuring construction quality

206. What challenges do robots face when recognizing and interpreting emotional cues in human-robot interaction for mental health support and therapy?

Answer: Recognizing and interpreting emotional cues in human-robot interaction for mental health support and therapy pose challenges. Robots must develop emotional intelligence and adapt their responses to provide effective emotional support.

207. How can robots use image-based gesture recognition in automotive safety systems for driver monitoring and accident prevention?

Answer: Robots apply image-based gesture recognition in automotive safety systems for driver monitoring and accident prevention. Recognizing driver gestures helps ensure safe driving by detecting distracted or fatigued drivers.

208. What is the significance of image processing in satellite-based remote sensing for agriculture, forestry, and natural resource management?

Answer: Image processing is significant in satellite-based remote sensing for agriculture, forestry, and natural resource management. It assists in monitoring land use, assessing environmental conditions, and optimizing resource management.

209. How do robots employ visual analysis in archaeological excavations for uncovering historical artifacts and preserving cultural heritage?

Answer: Robots employ visual analysis in archaeological excavations for uncovering historical artifacts and preserving cultural heritage. Visual data aids archaeologists in documenting and studying ancient civilizations.

210. What are the applications of facial recognition technology in healthcare for patient identification, medical record access, and security?

Answer: Facial recognition technology has applications in healthcare for patient identification, medical record access, and security. It ensures accurate patient records and enhances access control in healthcare facilities.

211. How can robots utilize visual tracking and recognition in the automotive industry for quality control, defect detection, and assembly line efficiency?

Answer: Robots use visual tracking and recognition in the automotive industry for quality control, defect detection, and assembly line efficiency. Visual data helps maintain manufacturing quality standards and optimize production processes.

212. What is the role of image recognition in augmented reality for creating immersive educational experiences and interactive museum exhibits?

Answer: Image recognition plays a vital role in augmented reality for creating immersive educational experiences and interactive museum exhibits. It enhances learning and engagement by allowing users to interact with virtual content in real-world environments.

213. How do robots employ visual perception in marine biology research for studying underwater ecosystems and marine biodiversity?

Answer: Robots utilize visual perception in marine biology research for studying underwater ecosystems and marine biodiversity. Cameras and visual data analysis aid scientists in documenting and understanding aquatic life.

214. What challenges do robots face when recognizing and interpreting visual and auditory cues in human-robot interaction for language translation and cross-cultural communication?

Answer: Recognizing and interpreting visual and auditory cues in human-robot interaction for language translation and cross-cultural communication pose challenges. Robots must bridge language barriers and adapt to cultural differences to facilitate effective communication.

215. How can robots apply image forgery detection in digital marketing and advertising to prevent fraudulent campaigns and maintain brand trust?

Answer: Robots apply image forgery detection in digital marketing and advertising to prevent fraudulent campaigns and maintain brand trust. It ensures that marketing materials are authentic and align with brand values.

216. What is the significance of image-based localization and navigation in robotics for precise positioning and mapping in complex environments?

Answer: Image-based localization and navigation are significant in robotics for precise positioning and mapping in complex environments. They enhance robot mobility and support tasks in challenging and unstructured spaces.

217. How do robots employ visual tracking for wildlife conservation to monitor animal behavior, migration patterns, and habitat changes?

Answer: Robots use visual tracking for wildlife conservation to monitor animal behavior, migration patterns, and habitat changes. Visual data contributes to scientific research and the preservation of ecosystems.

218. What are the benefits of using visual analytics in transportation and logistics for route optimization, fleet management, and supply chain visibility?

Answer: Visual analytics in transportation and logistics offer benefits by optimizing routes, managing fleets, and improving supply chain visibility. It enhances efficiency and reduces operational costs in the transportation industry.

219. How can robots utilize visual analysis in space exploration for planetary surface mapping, extraterrestrial research, and autonomous rover navigation?

Answer: Robots apply visual analysis in space exploration for planetary surface mapping, extraterrestrial research, and autonomous rover navigation. Visual data aids in exploring and studying celestial bodies, contributing to our understanding of the universe.

220. What role does image processing play in autonomous navigation for ground robots in agricultural field operations, such as planting and harvesting?

Answer: Image processing plays a crucial role in autonomous navigation for ground robots in agricultural field operations, including planting and harvesting. It supports precision agriculture by guiding robots in crop-related tasks.

221. How do robots employ visual perception in architecture and urban planning for designing sustainable and livable cities?

Answer: Robots utilize visual perception in architecture and urban planning for designing sustainable and livable cities. Visual data helps architects and planners assess environmental impact and create urban spaces that prioritize human well-being.

222. What challenges do robots face when interpreting and responding to visual and tactile cues in human-robot interaction for healthcare and rehabilitation applications?

Answer: Interpreting and responding to visual and tactile cues in human-robot interaction for healthcare and rehabilitation applications pose challenges. Robots must develop sensitivity to touch and adapt their responses to assist with medical procedures and therapy.

223. How can robots apply image-based gesture recognition in assistive technology for individuals with disabilities, enabling them to control devices and communicate effectively?

Answer: Robots use image-based gesture recognition in assistive technology for individuals with disabilities, enabling them to control devices and communicate effectively. Recognizing gestures empowers individuals with greater independence and accessibility.

224. What is the significance of image forgery detection in digital entertainment and gaming for preserving fair play and ensuring a level playing field?

Answer: Image forgery detection is significant in digital entertainment and gaming for preserving fair play and ensuring a level playing field. It helps prevent cheating and maintain the integrity of online gaming environments.

225. How can robots employ visual tracking and recognition in the construction industry for site progress monitoring, safety inspections, and project management?

Answer: Robots use visual tracking and recognition in the construction industry for site progress monitoring, safety inspections, and project management. Visual data aids in assessing construction quality, ensuring safety compliance, and managing project timelines.

226. What are the benefits of using visual analytics in energy production and distribution for optimizing grid operations, renewable energy integration, and resource management?
Answer: Visual analytics in energy production and distribution offer benefits by optimizing grid operations, integrating renewable energy sources, and managing resources efficiently. It supports the transition to sustainable and reliable energy systems.
227. How do robots utilize visual perception in underwater exploration for discovering shipwrecks, geological features, and underwater ecosystems?
Answer: Robots utilize visual perception in underwater exploration for discovering shipwrecks, geological features, and underwater ecosystems. Cameras and visual data analysis contribute to marine archaeology and environmental research.
228. What role does image recognition play in augmented reality for enhancing real-time information overlay and navigation assistance in various industries?
Answer: Image recognition plays a role in augmented reality by enhancing real-time information overlay and navigation assistance across various industries. It enables users to access contextual information and aids in navigation and decision-making.
229. How can robots apply image processing in agriculture for plant disease diagnosis, precision spraying, and yield prediction?
Answer: Robots use image processing in agriculture for plant disease diagnosis, precision spraying, and yield prediction. Visual data helps farmers optimize crop health, minimize pesticide use, and estimate crop yields accurately.
230. What is the role of image forgery detection in law enforcement and criminal investigations for identifying manipulated or counterfeit evidence?
Answer: Image forgery detection plays a crucial role in law enforcement and criminal investigations by identifying manipulated or counterfeit evidence. It ensures the integrity of visual materials used in legal proceedings.

ONLINE RESOURCES

Practice Paper Link: https://forms.gle/oCWPLTNtgxp1JQNf7

QR Code:

Video Content Link:
https://www.youtube.com/playlist?list=PLzlh_31VWEFpeQTLSEc9glTEiRST01xV-

QR Code:

Chapter 8: Robot Control Systems

1. How does open-loop control differ from closed-loop control in robot systems?
Answer: Open-loop control relies on predefined commands without feedback, while closed-loop control adjusts actions based on sensor feedback for precise and adaptive movements.
2. What is the purpose of a PID controller in robot control systems?
Answer: A PID controller regulates a robot's motion by adjusting proportional, integral, and derivative components to maintain accurate positions or trajectories.
3. How do robots achieve coordinated motion in multi-axis systems?
Answer: Robots achieve coordinated motion by synchronizing multiple axes or joints, ensuring precise and synchronized movement.
4. What is the role of kinematic equations in robot control?
Answer: Kinematic equations relate joint angles to end-effector positions, allowing robots to calculate and control their movements accurately.
5. What are the advantages of using inverse kinematics in robotics?
Answer: Inverse kinematics enables robots to determine joint angles required to reach a specific end-effector position, facilitating tasks like path planning and obstacle avoidance.
6. How do force and torque sensors contribute to robot control?
Answer: Force and torque sensors provide feedback on interaction forces, allowing robots to adapt their actions in response to external forces, ensuring safe and precise operations.
7. What is impedance control, and how does it benefit robot manipulation tasks?
Answer: Impedance control regulates a robot's response to external forces, making it adaptable for tasks like compliant assembly and interaction with soft or fragile objects.
8. What is the significance of trajectory planning in robot control systems?
Answer: Trajectory planning defines the desired path for a robot's end-effector, optimizing motion efficiency and ensuring smooth and collision-free movements.
9. How can robots utilize path following algorithms for navigation and exploration?
Answer: Robots use path following algorithms to follow predefined paths accurately, making them suitable for tasks like autonomous exploration and surveillance.
10. How does visual serving enhance robot control in tasks requiring visual feedback?
Answer: Visual serving uses visual cues to adjust a robot's position or orientation, enabling precise and adaptable actions in tasks like pick-and-place operations.
11. What role does adaptive control play in robot systems, particularly in changing environments?
Answer: Adaptive control allows robots to continuously adjust their control parameters based on changing conditions, enhancing performance and robustness in dynamic environments.
12. How do learning-based control approaches, such as reinforcement learning, improve robot autonomy and decision-making?
Answer: Learning-based control methods, like reinforcement learning, enable robots to learn optimal actions through trial and error, enhancing their autonomy and adaptability.
13. What are the challenges of real-time control in robotics, and how are they addressed?
Answer: Real-time control in robotics faces challenges related to computational speed and latency. These challenges are addressed through optimized algorithms and hardware.
14. What is the role of safety interlocks and emergency stop systems in robot control?
Answer: Safety interlocks and emergency stop systems provide mechanisms to halt robot operations in case of emergencies or unforeseen situations, ensuring human safety.
15. How does collaborative robot control enable safe interaction between robots and humans in shared workspaces?
Answer: Collaborative robot control includes safety features and compliance mechanisms that allow robots to work alongside humans safely, enabling human-robot collaboration.

16. How do mobile robot control systems navigate complex environments autonomously?
Answer: Mobile robot control systems employ sensors and algorithms for obstacle detection and avoidance, enabling autonomous navigation in various environments.
17. What is the importance of motion planning algorithms in robot control for collision avoidance and efficient path selection?
Answer: Motion planning algorithms ensure that robots navigate efficiently and avoid collisions, optimizing their movements in complex environments.
18. How do haptic feedback systems enhance teleoperation and remote control of robots?
Answer: Haptic feedback systems provide operators with tactile sensations, improving their control and perception when remotely operating robots.
19. What role do sensor fusion techniques play in robot control for combining data from multiple sensors?
Answer: Sensor fusion techniques integrate data from various sensors, enhancing the accuracy and reliability of robot control systems.
20. How can robots use machine learning algorithms for adaptive control and task optimization?
Answer: Machine learning algorithms allow robots to adapt to changing conditions and optimize their control strategies for improved performance.
21. What is the significance of self-calibration in robot control systems for maintaining accuracy over time?
Answer: Self-calibration processes help robots maintain their accuracy by periodically adjusting sensor parameters and eliminating calibration errors.
22. How do robots achieve real-time feedback control during high-speed manufacturing processes?
Answer: Robots achieve real-time feedback control through fast sensors and control algorithms to ensure precise and efficient operations in manufacturing.
23. How can robots apply predictive control strategies for anticipating and responding to future events or disturbances?
Answer: Predictive control strategies use models to anticipate future states and disturbances, allowing robots to proactively adjust their actions.
24. What role do redundancy resolution algorithms play in multi-arm robot control?
Answer: Redundancy resolution algorithms optimize the coordination of multiple robot arms, ensuring efficient and collision-free movements.
25. How can robots implement energy-efficient control strategies for sustainability in various applications?
Answer: Energy-efficient control strategies optimize robot movements and reduce power consumption, contributing to sustainability in robotic applications.
26. What challenges are associated with decentralized control systems in multi-robot teams, and how are they addressed?
Answer: Decentralized control systems face challenges related to coordination and communication among robots. These challenges are addressed through robust communication protocols and coordination algorithms.
27. How do robots apply neural network-based controllers for complex tasks and decision-making?
Answer: Neural network-based controllers enable robots to handle complex tasks and make decisions by learning from data and adapting their control strategies.
28. What is the role of fault tolerance mechanisms in robot control systems for maintaining operation in the presence of failures?
Answer: Fault tolerance mechanisms allow robots to continue operating even when certain components or sensors fail, ensuring system reliability.
29. How can robots implement adaptive control strategies for improving performance in uncertain environments?
Answer: Adaptive control strategies enable robots to continuously adjust their parameters based on environmental changes, enhancing performance and robustness.
30. How does real-time feedback control in robotic surgery ensure precision and patient safety during procedures?

Answer: Real-time feedback control in robotic surgery ensures precise instrument movements and enhances patient safety by minimizing errors and tremors.

31. What is the significance of compliance control in robots used for tasks like material handling and assembly?

Answer: Compliance control allows robots to adjust their stiffness and compliance, making them adaptable for tasks that require interaction with delicate or deformable objects.

32. How do robots apply model predictive control for optimizing trajectories in dynamic environments?

Answer: Model predictive control uses predictive models to optimize robot trajectories, ensuring efficient and safe movements in dynamic environments.

33. How can robots use reinforcement learning for autonomous decision-making in navigation and exploration tasks?

Answer: Reinforcement learning allows robots to learn optimal navigation and exploration strategies through trial and error, improving their autonomy.

34. What role do advanced control algorithms, such as fuzzy logic and genetic algorithms, play in robot control systems?

Answer: Advanced control algorithms like fuzzy logic and genetic algorithms offer enhanced control strategies for robots, particularly in complex and non-linear tasks.

35. How does multi-modal sensory integration contribute to robot perception and control in human-robot interaction scenarios?

Answer: Multi-modal sensory integration combines data from various sensors, improving robot perception and enabling effective human-robot interaction.

36. What is the impact of real-time data processing and control in autonomous vehicles for safe and efficient transportation?

Answer: Real-time data processing and control are critical for autonomous vehicles to make split-second decisions, ensuring safe and efficient transportation.

37. How can robots leverage probabilistic control strategies for uncertainty management in navigation and decision-making?

Answer: Probabilistic control strategies allow robots to account for uncertainty in sensor data and make probabilistic decisions, enhancing robustness in uncertain environments.

38. What challenges do swarm robotics systems face in decentralized control and coordination, and how are they overcome?

Answer: Swarm robotics systems face challenges in decentralized control and coordination, which are addressed through swarm algorithms and communication protocols.

39. How does trajectory tracking control improve the accuracy of robots in tasks like 3D printing and CNC machining?

Answer: Trajectory tracking control ensures that robots follow precise paths, enhancing their accuracy and performance in tasks like 3D printing and CNC machining.

40. What is the role of compliance control in robotic exoskeletons for assisting individuals with mobility impairments?

Answer: Compliance control in robotic exoskeletons allows them to adapt to users' movements and assist individuals with mobility impairments, improving their quality of life.

41. How do robots implement predictive maintenance strategies, and what benefits do they offer in industrial automation?

Answer: Robots employ predictive maintenance by analyzing sensor data to predict equipment failures and schedule maintenance before breakdowns occur. This approach minimizes downtime, reduces maintenance costs, and extends the lifespan of industrial machinery.

42. What is the role of motion control systems in high-precision applications like semiconductor manufacturing and micro-assembly?

Answer: Motion control systems provide precise control of robotic movements in applications requiring nanometer-level accuracy, such as semiconductor manufacturing and micro-assembly. They ensure the quality and reliability of these delicate processes.

43. How can robots utilize reinforcement learning to adapt to changing environments in dynamic tasks like autonomous exploration and disaster response?

Answer: Reinforcement learning enables robots to learn optimal strategies through interaction with their environments, making them adaptable in dynamic scenarios like autonomous exploration and disaster response, where conditions may change rapidly.

44. What is the significance of energy-aware control strategies in autonomous mobile robots for extending operational endurance in remote or resource-constrained environments?

Answer: Energy-aware control strategies optimize a robot's operations to maximize its battery life or fuel efficiency, crucial for extending operational endurance in remote or resource-constrained environments.

45. How do robots apply real-time vision-based control for tasks like object tracking and manipulation in manufacturing and logistics?

Answer: Real-time vision-based control uses cameras and image processing to guide robots in tasks like object tracking and manipulation, enhancing efficiency and accuracy in manufacturing and logistics.

46. What challenges arise in multi-robot coordination, and how can decentralized control systems address these challenges?

Answer: Multi-robot coordination faces challenges in ensuring cooperation and avoiding collisions. Decentralized control systems leverage local communication and decision-making to address these challenges while minimizing centralized control.

47. How do advanced control algorithms like neural networks enhance the capabilities of robotic drones for tasks like autonomous flight and aerial photography?

Answer: Advanced control algorithms like neural networks enable robotic drones to perform complex tasks such as autonomous flight and aerial photography by learning from data and adapting to changing conditions.

48. What is the role of tactile feedback systems in robot-assisted surgery, and how do they improve surgical precision and safety?

Answer: Tactile feedback systems provide surgeons with a sense of touch during robot-assisted surgery, enhancing precision and safety by allowing them to feel tissue characteristics and pressure.

49. How can robots implement adaptive control for manufacturing processes that involve variable materials, tool wear, and environmental factors?

Answer: Adaptive control allows robots to adjust their parameters and strategies in real-time to account for variable materials, tool wear, and environmental conditions, ensuring consistent manufacturing quality.

50. What is the significance of end-effector compliance control in tasks requiring delicate interaction with objects, such as fruit harvesting or medical procedures?

Answer: End-effector compliance control makes robots adaptable to tasks like fruit harvesting or medical procedures, allowing them to interact delicately with objects by adjusting stiffness and compliance.

51. How do robots utilize distributed control algorithms in swarm robotics to achieve coordinated behavior without centralized coordination?

Answer: Distributed control algorithms enable robots in swarm robotics to communicate and coordinate their actions locally, achieving collective behavior without the need for centralized coordination.

52. How does motion planning with dynamic obstacles enhance the safety and efficiency of autonomous vehicles in urban environments?

Answer: Motion planning with dynamic obstacles allows autonomous vehicles to adapt to changing traffic conditions and unexpected obstacles, improving safety and efficiency in urban environments.

53. What role does model-based control play in industrial robots for maintaining accuracy and consistency in repetitive tasks like welding or painting?

Answer: Model-based control uses mathematical models to maintain the accuracy and consistency of industrial robots in repetitive tasks like welding or painting, reducing errors and improving quality.

54. How can robots apply impedance control to interact safely with humans in collaborative settings like rehabilitation therapy or healthcare assistance?

Answer: Impedance control allows robots to adjust their stiffness and compliance to interact safely with humans in collaborative settings, such as rehabilitation therapy or healthcare assistance, preventing injuries.

55. What challenges do robots face in underwater control systems, and how are they overcome to enable tasks like marine research and exploration?

Answer: Underwater control systems face challenges related to communication delays and buoyancy control. These challenges are addressed through specialized algorithms and robust hardware to enable tasks like marine research and exploration.

56. How does visual-inertial sensor fusion improve robot perception and navigation in environments with limited GPS coverage, such as indoor spaces or urban canyons?

Answer: Visual-inertial sensor fusion combines visual and inertial sensor data to enhance robot perception and navigation in environments with limited GPS coverage, enabling accurate localization and mapping.

57. What is the role of learning-based control in soft robotics for tasks like soft gripper manipulation and compliant locomotion?

Answer: Learning-based control in soft robotics allows robots to adapt their behaviors and actions based on the properties of soft materials, making them suitable for tasks like soft gripper manipulation and compliant locomotion.

58. How do robots apply hybrid control strategies that combine both feedback and feedforward control to achieve precise and efficient movements?

Answer: Hybrid control strategies combine feedback from sensors with pre-planned feedforward control to achieve precise and efficient robot movements in tasks that require high accuracy.

59. What is the significance of explainable AI in robot control systems, particularly in applications where human trust and understanding are crucial?

Answer: Explainable AI in robot control systems provides transparency and interpretable decision-making, enhancing human trust and understanding, especially in applications like autonomous vehicles and healthcare.

60. How can robots implement reinforcement learning for adaptive grasping and manipulation of objects with varying shapes and properties?

Answer: Reinforcement learning enables robots to learn optimal grasping and manipulation strategies for objects with varying shapes and properties, making them versatile in tasks like pick-and-place operations.

61. What role do haptic sensors play in teleoperation and remote control of robots for tasks like bomb disposal or space exploration?

Answer: Haptic sensors provide feedback to remote operators, allowing them to feel the environment and interact with objects during tasks like bomb disposal or space exploration, enhancing control and safety.

62. How can robots apply event-based control strategies for tasks that require rapid response times, such as autonomous racing or drone racing?

Answer: Event-based control strategies allow robots to respond rapidly to events or changes in the environment, making them suitable for tasks like autonomous racing or drone racing that demand quick reactions.

63. What is the impact of real-time simulation in robot control system development for testing and optimizing control algorithms?

Answer: Real-time simulation enables developers to test and optimize control algorithms in a virtual environment before deploying them on physical robots, saving time and resources.

64. How do robots utilize bio-inspired control strategies for tasks that mimic animal behaviors, such as swarm intelligence or quadruped locomotion?

Answer: Bio-inspired control strategies mimic animal behaviors, allowing robots to perform tasks like swarm intelligence or quadruped locomotion more efficiently and adaptively.

65. What are the challenges and advantages of distributed control systems in space exploration missions using multiple robotic spacecraft?

Answer: Distributed control systems in space exploration missions face challenges in communication and coordination. However, they offer redundancy, fault tolerance, and increased mission flexibility when using multiple robotic spacecraft.

66. How do robots employ real-time kinematic (RTK) GPS in precision agriculture for tasks like autonomous crop spraying and soil mapping?

Answer: Robots use RTK GPS technology in precision agriculture to achieve centimeter-level positioning accuracy. This enables them to perform tasks like autonomous crop spraying with high precision and create detailed soil maps for optimized farming practices.

67. What is the role of adaptive control in robotic exoskeletons for enhancing the mobility and rehabilitation of individuals with mobility impairments?

Answer: Adaptive control in robotic exoskeletons enables personalized assistance by adjusting to users' needs and abilities, enhancing mobility and rehabilitation for individuals with mobility impairments.

68. How can robots implement hybrid force-motion control for tasks like polishing and deburring, which require both force and positional control?

Answer: Hybrid force-motion control combines force and positional control, allowing robots to perform tasks like polishing and deburring with the necessary precision while maintaining force compliance.

69. What challenges arise in underwater robot control systems, and how are they addressed for tasks like pipeline inspection and ocean exploration?

Answer: Underwater robot control systems face challenges related to buoyancy control and communication. These challenges are addressed through specialized control algorithms and robust underwater communication systems for tasks like pipeline inspection and ocean exploration.

70. How do robots apply reinforcement learning in autonomous vehicles to enhance decision-making and adapt to complex traffic scenarios?

Answer: Reinforcement learning in autonomous vehicles enhances decision-making by allowing them to learn from various traffic scenarios, adapting to complex situations and improving safety on the road.

71. What is the significance of soft robotics control for applications like assistive devices and human-robot interaction?

Answer: Soft robotics control enables robots to interact safely and adapt to the human body's soft and deformable nature, making them suitable for assistive devices and enhancing human-robot interaction.

72. How can robots employ event-based control for tasks requiring rapid and precise responses, such as drone racing or robotic surgery?

Answer: Event-based control strategies enable robots to respond rapidly to specific events or changes in the environment, making them suitable for tasks demanding quick and precise reactions.

73. What challenges do swarm robotics systems face in distributed control and coordination, and how are they overcome for applications like search and rescue?

Answer: Swarm robotics systems face challenges in communication and coordination among a large number of robots. These challenges are addressed through robust swarm algorithms and decentralized control for applications like search and rescue.

74. How does teleoperation with haptic feedback enhance robot control in scenarios like deep-sea exploration or hazardous environments?

Answer: Teleoperation with haptic feedback allows operators to feel and interact with the remote environment, improving robot control in scenarios where human presence is risky or impossible, such as deep-sea exploration or hazardous environments.

75. What is the impact of cognitive control in humanoid robots for tasks like human-like locomotion and social interaction?

Answer: Cognitive control in humanoid robots enables human-like locomotion and social interaction by simulating human decision-making processes and adapting to dynamic environments.

76. How can robots implement deep reinforcement learning for autonomous navigation and decision-making in complex environments like urban settings?

Answer: Deep reinforcement learning enables robots to learn optimal navigation and decision-making policies through extensive training, making them capable of autonomous operation in complex urban environments.

77. What role do hybrid control systems play in legged robots to achieve stability and adaptability in challenging terrains?

Answer: Hybrid control systems combine different control modes to achieve stability and adaptability in legged robots, allowing them to traverse challenging terrains and perform dynamic movements.

78. How do robots use impedance control for tasks like human-robot collaboration in manufacturing and healthcare?

Answer: Impedance control allows robots to adjust their stiffness and compliance to interact safely and effectively with humans in tasks such as manufacturing and healthcare assistance.

79. What challenges and benefits are associated with the integration of tactile sensors into robot control systems for tasks like object manipulation and surface exploration?

Answer: Integrating tactile sensors into robot control systems poses challenges in sensor data interpretation. However, it provides benefits by enhancing object manipulation and enabling surface exploration in unknown environments.

80. How can robots employ model predictive control for precise and efficient control in applications like autonomous racing and drone flight?

Answer: Model predictive control uses predictive models to optimize robot actions and achieve precise and efficient control, making it valuable in applications like autonomous racing and drone flight.

81. What is the role of explainable AI in enhancing the transparency and trustworthiness of robot decision-making in critical applications like healthcare and finance?

Answer: Explainable AI in robot control systems provides interpretable decision-making, increasing transparency and trustworthiness in critical applications such as healthcare and finance, where human lives and financial assets are at stake.

82. How do robots use reinforcement learning for adaptive grasping and manipulation of objects with varying shapes and sizes in logistics and warehouse automation?

Answer: Reinforcement learning enables robots to adapt their grasping and manipulation strategies to handle objects with varying shapes and sizes, improving efficiency in logistics and warehouse automation.

83. What challenges and advantages arise in the implementation of adaptive control systems in unmanned aerial vehicles (UAVs) for tasks like surveillance and reconnaissance?

Answer: Adaptive control systems in UAVs face challenges related to dynamic environments and varying payloads. However, they offer advantages such as improved stability and performance in surveillance and reconnaissance missions.

84. How can robots leverage event-based vision systems for tasks that require high-speed object detection and tracking, such as autonomous racing and sports analysis?

Answer: Event-based vision systems enable robots to detect and track objects at high speeds, making them suitable for tasks like autonomous racing and sports analysis, where rapid visual processing is essential.

85. What is the significance of reinforcement learning in robotics for tasks like robotic cooking, where the robot must learn complex recipes and adapt to different cooking environments?

Answer: Reinforcement learning allows robots to learn complex cooking tasks, adapt to various recipes, and optimize their actions in different cooking environments, enhancing their culinary capabilities.

86. How do robots apply compliant control in tasks like human-robot collaboration in industrial settings, ensuring safety and productivity?

Answer: Compliant control allows robots to adjust their stiffness and compliance to work safely alongside humans in industrial settings, improving safety and productivity in collaborative tasks.

87. What challenges and solutions are associated with the control of autonomous underwater vehicles (AUVs) for tasks like deep-sea exploration and oceanographic research?

Answer: AUV control faces challenges in communication and navigation in underwater environments. Solutions include acoustic communication and advanced navigation algorithms, enabling deep-sea exploration and research.

88. How can robots implement hybrid force-velocity control for tasks like polishing or grinding, where both force and velocity control are essential?

Answer: Hybrid force-velocity control combines force and velocity control modes, allowing robots to perform tasks like polishing and grinding with precision and efficiency while maintaining force compliance.

89. What is the impact of human-in-the-loop control in drones and robotic systems for applications like disaster response and environmental monitoring?

Answer: Human-in-the-loop control allows human operators to provide guidance and decision-making input to drones and robots, enhancing their effectiveness in tasks such as disaster response and environmental monitoring.

90. How do robots apply sensory feedback integration in tasks like tele-manipulation and remote control of robotic arms for tasks like bomb disposal of hazardous material handling?

Answer: Sensory feedback integration provides operators with real-time feedback from the robotic arm's sensors, improving control and safety in tasks like bomb disposal and hazardous material handling.

91. What challenges and benefits arise in the control of autonomous delivery robots for last-mile logistics and e-commerce applications?

Answer: Autonomous delivery robots face challenges related to navigation in urban environments. However, they offer benefits such as efficient last-mile delivery and reduced traffic congestion in e-commerce applications.

92. How can robots use bio-inspired control strategies for tasks like agile aerial maneuvering or walking on uneven terrain?

Answer: Bio-inspired control strategies mimic natural behaviors and movements, enabling robots to perform tasks like agile aerial maneuvering and walking on uneven terrain with greater agility and adaptability.

93. What role do emotion recognition and affective computing play in human-robot interaction and assistive robotics for healthcare and social support?

Answer: Emotion recognition and affective computing enable robots to understand and respond to human emotions, improving their effectiveness in healthcare and social support roles.

94. How can robots apply reinforcement learning for autonomous decision-making in agricultural automation, such as precision planting and harvesting?

Answer: Reinforcement learning allows robots to make autonomous decisions in agricultural tasks like precision planting and harvesting, optimizing crop yield and resource utilization.

95. What is the significance of sensor fusion and simultaneous localization and mapping (SLAM) in robot control for autonomous exploration and mapping of unknown environments?

Answer: Sensor fusion and SLAM techniques enable robots to create maps of unknown environments while localizing themselves accurately, enhancing autonomous exploration and mapping capabilities.

96. How do robots utilize predictive control strategies in human-robot interaction scenarios, such as physical therapy and rehabilitation?

Answer: Predictive control strategies anticipate human movements and adapt robot behavior accordingly in physical therapy and rehabilitation, enhancing therapy effectiveness and patient outcomes.

97. What challenges and benefits are associated with the control of soft robots for tasks like soft gripper manipulation and medical applications?

Answer: Soft robot control faces challenges in modeling and control algorithms. However, it offers benefits such as safe interaction with delicate objects and human tissues in medical applications.

98. How can robots apply reinforcement learning for adaptive navigation and exploration in challenging terrains, such as planetary exploration or search and rescue missions?

Answer: Reinforcement learning enables robots to learn adaptive navigation and exploration strategies in challenging terrains like planetary surfaces or disaster-stricken areas, improving their efficiency and effectiveness.

99. What is the role of explainable AI in autonomous vehicles for enhancing trust and safety in self-driving cars and drones?

Answer: Explainable AI in autonomous vehicles provides transparent decision-making, increasing trust and safety in self-driving cars and drones by allowing users to understand and validate system actions.

100. How does cognitive control enhance robot decision-making and problem-solving capabilities in applications like warehouse automation and logistics?

Answer: Cognitive control equips robots with problem-solving capabilities, allowing them to make intelligent decisions in dynamic environments, thereby improving efficiency in warehouse automation and logistics.

101. How do robots employ reinforcement learning in adaptive locomotion, allowing them to traverse challenging terrains like rocky landscapes or snow-covered areas?

Answer: Reinforcement learning enables robots to learn adaptive locomotion strategies through trial and error, allowing them to navigate and traverse challenging terrains with agility and stability.

102. What challenges and benefits are associated with swarm robotics control in applications like environmental monitoring and disaster response?

Answer: Swarm robotics control faces challenges in coordination and scalability but offers benefits such as improved resilience, flexibility, and coverage in environmental monitoring and disaster response scenarios.

103. How can robots use bio-inspired control strategies for tasks like mimicry of animal locomotion, such as snake-like crawling or bird-like flight?

Answer: Bio-inspired control strategies mimic animal locomotion patterns, enabling robots to perform tasks like snake-like crawling or bird-like flight with efficiency and adaptability.

104. What is the significance of model-based control in robotic prosthetics for providing natural and intuitive movements to individuals with limb loss?

Answer: Model-based control in robotic prosthetics uses mathematical models to replicate natural limb movements, enhancing mobility and providing a more intuitive experience for individuals with limb loss.

105. How do robots apply reinforcement learning for autonomous decision-making in underwater exploration and marine research missions?

Answer: Reinforcement learning allows robots to make autonomous decisions during underwater exploration and marine research missions, adapting to changing underwater conditions and optimizing data collection.

106. What role does motion prediction play in robot control for autonomous vehicles, particularly in scenarios involving pedestrian safety and collision avoidance?

Answer: Motion prediction algorithms help autonomous vehicles anticipate the movements of pedestrians and other vehicles, enabling safer and more effective collision avoidance strategies.

107. How can robots implement shared autonomy in scenarios like drone piloting, where human operators and automated systems collaborate for enhanced performance?

Answer: Shared autonomy in drone piloting combines human expertise with automated systems to enhance performance, allowing operators to focus on high-level tasks while automated systems handle low-level control.

108. What challenges and solutions are associated with the control of autonomous agricultural robots for tasks like precision weeding and harvesting?

Answer: Autonomous agricultural robots face challenges in plant recognition and navigation. Solutions include computer vision and GPS-based guidance systems for tasks like precision weeding and harvesting.

109. How do robots utilize emotion recognition in human-robot interaction to adapt their behavior and responses to human emotional states?

Answer: Emotion recognition enables robots to detect and respond to human emotional states, enhancing the quality of human-robot interaction and providing emotional support when needed.

110. What is the impact of reinforcement learning in robot-assisted surgery, where precision and safety are critical for patient outcomes?

Answer: Reinforcement learning in robot-assisted surgery allows robots to continuously improve their surgical precision and adapt to variations, ultimately enhancing patient outcomes and safety.

111. How can robots apply vision-based control in applications like autonomous inspection and quality control on manufacturing assembly lines?

Answer: Vision-based control relies on cameras and image processing to guide robots in tasks like inspection and quality control on manufacturing assembly lines, ensuring product quality and consistency.

112. What challenges and benefits arise in the control of swarm robots for tasks like environmental cleanup and agriculture?

Answer: Swarm robot control poses challenges in maintaining coordination and communication among a large number of robots. However, it offers benefits such as increased efficiency and coverage in tasks like environmental cleanup and agriculture.

113. How does adaptive control enhance the performance and efficiency of industrial robots in tasks like material handling and welding, especially in changing production environments?

Answer: Adaptive control enables industrial robots to adjust to changing production conditions, improving performance and efficiency in tasks like material handling and welding while maintaining high-quality standards.

114. How can robots implement multi-modal sensory integration for enhanced perception in applications like autonomous vehicles and assistive devices?

Answer: Multi-modal sensory integration combines data from various sensors, such as cameras and LiDAR, to improve perception and enable robust operation in autonomous vehicles and assistive devices.

115. What is the role of explainable AI in enhancing the interpretability and accountability of robot decision-making, particularly in legal and ethical contexts?

Answer: Explainable AI provides transparency in robot decision-making, making it accountable and interpretable in legal and ethical contexts, where understanding the basis of decisions is crucial.

116. How do robots use reinforcement learning in adaptive grasping and manipulation of objects with uncertain properties, such as irregularly shaped items or deformable materials?

Answer: Reinforcement learning enables robots to adapt their grasping and manipulation strategies to handle objects with uncertain properties, improving versatility in tasks involving irregularly shaped items or deformable materials.

117. What challenges and solutions are associated with the control of drones for precision agriculture, where tasks include crop monitoring and pesticide spraying?

Answer: Drone control in precision agriculture faces challenges in flight stability and automated navigation. Solutions include GPS guidance and obstacle avoidance systems for tasks like crop monitoring and pesticide spraying.

118. How can robots apply predictive control in logistics and warehouse automation for efficient inventory management and order fulfillment?

Answer: Predictive control in logistics and warehouse automation optimizes inventory management and order fulfillment by predicting demand and dynamically adjusting operations to meet customer needs.

119. What role do hybrid control strategies play in legged robots for achieving both stability and agility in dynamic environments?

Answer: Hybrid control strategies balance stability and agility in legged robots by switching between control modes as needed, allowing them to navigate challenging and dynamic environments effectively.

120. How does cognitive control enhance robots' adaptability in human-robot collaboration scenarios, such as manufacturing and assembly tasks?

Answer: Cognitive control enables robots to adapt their actions and decisions based on the context of human-robot collaboration, improving efficiency and safety in tasks like manufacturing and assembly.

121. How can robots implement energy-aware control strategies for sustainability in applications like energy-efficient HVAC systems and autonomous electric vehicles?

Answer: Energy-aware control strategies optimize the operation of robots, such as HVAC systems and electric vehicles, to reduce energy consumption, enhance sustainability, and extend battery life.

122. What challenges and benefits arise in the control of robotic exoskeletons for rehabilitation and mobility assistance, particularly for individuals with spinal cord injuries?

Answer: Robotic exoskeleton control poses challenges related to user-specific customization. However, it offers benefits like improved mobility and quality of life for individuals with spinal cord injuries.

123. How do robots use sensory feedback integration for safe human-robot interaction in shared workspaces, such as collaborative manufacturing environments?

Answer: Sensory feedback integration ensures safe human-robot interaction in shared workspaces by providing real-time feedback on proximity and touch, allowing robots to adjust their actions accordingly.

124. What is the significance of reinforcement learning in autonomous vehicles for improving adaptive driving behavior and traffic management?

Answer: Reinforcement learning in autonomous vehicles enhances adaptive driving behavior, making them better at handling diverse traffic scenarios and contributing to more efficient traffic management.

125. How can robots apply model predictive control for precision alignment and calibration tasks in semiconductor manufacturing and optical systems?

Answer: Model predictive control ensures precision alignment and calibration in tasks like semiconductor manufacturing and optical systems by optimizing control actions to achieve high accuracy.

126. What role does event-based control play in drones for aerial photography and surveillance, where fast and precise movements are crucial?

Answer: Event-based control enables drones to respond rapidly to changes in the environment, making them ideal for tasks like aerial photography and surveillance that require fast and precise movements.

127. What challenges and solutions are associated with the control of autonomous underwater gliders for oceanographic research and data collection in remote underwater regions?

Answer: Autonomous underwater gliders face challenges related to communication and navigation in remote underwater regions. Solutions include acoustic communication and advanced navigation algorithms for oceanographic research and data collection.

128. How do robots employ adaptive control in surgical robots for tasks like delicate tissue manipulation and microsurgery?

Answer: Adaptive control in surgical robots allows them to adapt to the varying properties of tissues during delicate procedures, improving precision and safety in tasks like microsurgery.

129. What is the impact of real-time simulation in robot control system development for testing and validating control algorithms in complex and dynamic environments?

Answer: Real-time simulation accelerates the development of robot control systems by providing a virtual environment for testing and validating control algorithms in complex and dynamic scenarios.

130. How can robots implement impedance control for safe and precise interactions in applications like contact force control and object manipulation in unstructured environments?

Answer: Impedance control allows robots to adjust their stiffness and compliance, ensuring safe and precise interactions during tasks like contact force control and object manipulation in unstructured environments.

131. How do robots use model-based control for precise and stable flight in unmanned aerial vehicles (UAVs) during tasks like aerial mapping and surveillance?

Answer: Model-based control in UAVs employs mathematical models to achieve precise and stable flight, enhancing the accuracy of tasks like aerial mapping and surveillance.

132. What challenges and benefits arise in the control of robots for underwater archaeology and exploration missions in deep-sea environments?

Answer: Robot control in deep-sea environments faces challenges in pressure resistance and communication. Benefits include the ability to conduct detailed archaeological and exploration missions in deep-sea regions.

133. How can robots apply predictive maintenance strategies in industrial automation for minimizing downtime and optimizing production efficiency?

Answer: Robots employ predictive maintenance by analyzing sensor data to predict equipment failures, allowing timely maintenance to minimize downtime and optimize production efficiency in industrial automation.

134. What is the role of adaptive control in autonomous vehicles for enhancing safety and handling in various driving conditions?

Answer: Adaptive control in autonomous vehicles adjusts control parameters based on road conditions and vehicle dynamics, improving safety and handling in diverse driving conditions.

135. How do robots utilize multi-robot coordination algorithms for tasks like warehouse automation and collaborative search and rescue missions?

Answer: Multi-robot coordination algorithms enable robots to work together efficiently, enhancing warehouse automation and enabling collaborative search and rescue missions by optimizing resource utilization.

136. What is the significance of compliance control in human-robot collaboration for tasks like precision surgery and construction?

Answer: Compliance control allows robots to adjust their stiffness and compliance, ensuring safe human-robot collaboration in tasks like precision surgery and construction.

137. How can robots apply reinforcement learning for adaptive locomotion in legged robots, allowing them to traverse complex terrains and obstacles?

Answer: Reinforcement learning enables legged robots to adapt their locomotion strategies and navigate complex terrains and obstacles with agility and stability.

138. What challenges and solutions are associated with the control of robots for agricultural automation in tasks like fruit picking and autonomous tractors?

Answer: Agricultural robot control faces challenges in perception and navigation. Solutions include computer vision and GPS-based guidance systems for tasks like fruit picking and autonomous tractors.

139. How do robots employ vision-based control for autonomous underwater vehicles (AUVs) to perform tasks like underwater mapping and exploration?

Answer: Vision-based control in AUVs uses cameras and image processing to guide their actions in underwater tasks like mapping and exploration, enhancing their capabilities in underwater environments.

140. What is the role of explainable AI in enhancing the trust and interpretability of robot decision-making in applications like financial trading and autonomous trading systems?

Answer: Explainable AI in robot control systems provides transparent decision-making, increasing trust and interpretability in applications like financial trading and autonomous trading systems, where decisions can have significant financial impacts.

141. How can robots utilize sensory feedback integration for tasks like teleoperation and remote control of robotic arms in hazardous environments like nuclear facilities or chemical plants?

Answer: Sensory feedback integration allows operators to receive real-time feedback from the robot's sensors during teleoperation, enhancing control and safety in hazardous environments like nuclear facilities and chemical plants.

142. What challenges and benefits arise in the control of robotic drones for tasks like environmental monitoring and wildlife conservation in remote or rugged terrain?

Answer: Drone control in remote or rugged terrain poses challenges in navigation and communication. Benefits include efficient environmental monitoring and wildlife conservation efforts in challenging locations.

143. How does adaptive control enhance the performance of soft robots for tasks like delicate object handling and exploration in confined spaces?

Answer: Adaptive control in soft robots enables them to adjust their behaviors for tasks like delicate object handling and exploration in confined spaces, improving versatility and adaptability.

144. What is the impact of event-based control in robot soccer and robotic sports, where rapid and dynamic movements are essential?

Answer: Event-based control allows robots in robot soccer and robotic sports to respond rapidly to dynamic events and changes in the game, enhancing their competitiveness and agility.

145. How can robots implement energy-efficient control strategies in autonomous electric vehicles for extended range and reduced environmental impact?

Answer: Energy-efficient control strategies optimize the operation of autonomous electric vehicles, extending their range and reducing their environmental impact by conserving energy.

146. What role does cognitive control play in humanoid robots for tasks like natural language understanding and social interaction in service and healthcare settings?

Answer: Cognitive control equips humanoid robots with the ability to understand natural language and engage in social interaction, enhancing their effectiveness in service and healthcare roles.

147. How do robots use reinforcement learning for autonomous exploration and mapping of underground environments, such as mines and caves?
Answer: Reinforcement learning enables robots to autonomously explore and map underground environments like mines and caves, adapting to dynamic conditions and improving safety.
148. What challenges and solutions are associated with the control of swarm robots for precision agriculture, where tasks include crop monitoring and precision spraying?
Answer: Swarm robot control in precision agriculture faces challenges in communication and coordination. Solutions include swarm algorithms and distributed control for tasks like crop monitoring and precision spraying.
149. How can robots apply predictive control in robotic surgery for precise and stable movements during procedures like laparoscopic surgery and microsurgery?
Answer: Predictive control in robotic surgery uses predictive models to ensure precise and stable movements during procedures like laparoscopic surgery and microsurgery, improving surgical outcomes.
150. What is the significance of explainable AI in robot-assisted education and tutoring, where transparency and feedback are critical for effective learning?
Answer: Explainable AI in robot-assisted education provides transparent explanations and feedback to learners, enhancing their understanding and improving the effectiveness of educational robots.
151. How do robots employ real-time simulation in the development of autonomous drones for testing and optimizing navigation and control algorithms?
Answer: Real-time simulation allows developers to test and optimize navigation and control algorithms for autonomous drones in a virtual environment, accelerating development and ensuring safety.
152. What role does tactile sensing and haptic feedback play in robot control for tasks like braille writing and object recognition for visually impaired individuals?
Answer: Tactile sensing and haptic feedback enable robots to perform tasks like braille writing and object recognition for visually impaired individuals, enhancing accessibility and independence.
153. How can robots use impedance control for precise and compliant interactions in applications like rehabilitation therapy and physical assistance for the elderly?
Answer: Impedance control allows robots to provide precise and compliant interactions in applications like rehabilitation therapy and physical assistance for the elderly, improving safety and comfort.
154. What challenges and benefits arise in the control of autonomous drones for aerial surveying and mapping in remote and challenging terrain?
Answer: Autonomous drone control for aerial surveying and mapping faces challenges in navigation and endurance. Benefits include efficient data collection and mapping in remote and challenging terrains.
155. How does sensory fusion and sensor calibration enhance robot control in applications like autonomous vehicles and industrial automation?
Answer: Sensory fusion combines data from various sensors and sensor calibration ensures accuracy, enhancing robot control in applications like autonomous vehicles and industrial automation by providing reliable and precise information.
156. What is the role of reinforcement learning in robot-assisted therapy for children with autism, where personalized and adaptive interaction is crucial?
Answer: Reinforcement learning enables robots to provide personalized and adaptive interaction in robot-assisted therapy for children with autism, helping them develop social and communication skills.
157. How can robots implement hybrid force-impedance control for tasks like fine assembly and delicate object manipulation in manufacturing and electronics assembly?
Answer: Hybrid force-impedance control combines force and impedance control to perform tasks like fine assembly and delicate object manipulation with precision and sensitivity, ensuring quality in manufacturing and electronics assembly.
158. What challenges and solutions are associated with the control of aerial drones for wildlife monitoring and conservation, especially in remote and inaccessible areas?

Answer: Aerial drone control for wildlife monitoring faces challenges in navigation and battery life. Solutions include advanced navigation algorithms and efficient energy management for conservation efforts in remote and inaccessible areas.

159. How do robots employ multi-modal interaction in human-robot collaboration scenarios, enhancing communication and teamwork in industrial settings?

Answer: Multi-modal interaction enables robots to communicate through various modalities like speech, gestures, and touch, enhancing communication and teamwork in industrial settings for improved efficiency.

160. What is the impact of cognitive control in swarm robotics for tasks like disaster response and environmental monitoring, where adaptive decision-making is critical?

Answer: Cognitive control in swarm robotics enables adaptive decision-making in tasks like disaster response and environmental monitoring, improving the efficiency and effectiveness of collective actions.

161. How do robots use reinforcement learning for autonomous exploration and mapping of underground environments, such as mines and caves?

Answer: Reinforcement learning enables robots to autonomously explore and map underground environments like mines and caves, adapting to dynamic conditions and improving safety.

162. What challenges and solutions are associated with the control of swarm robots for precision agriculture, where tasks include crop monitoring and precision spraying?

Answer: Swarm robot control in precision agriculture faces challenges in communication and coordination. Solutions include swarm algorithms and distributed control for tasks like crop monitoring and precision spraying.

163. How can robots apply predictive control in robotic surgery for precise and stable movements during procedures like laparoscopic surgery and microsurgery?

Answer: Predictive control in robotic surgery uses predictive models to ensure precise and stable movements during procedures like laparoscopic surgery and microsurgery, improving surgical outcomes.

164. What is the significance of explainable AI in robot-assisted education and tutoring, where transparency and feedback are critical for effective learning?

Answer: Explainable AI in robot-assisted education provides transparent explanations and feedback to learners, enhancing their understanding and improving the effectiveness of educational robots.

165. How do robots employ real-time simulation in the development of autonomous drones for testing and optimizing navigation and control algorithms?

Answer: Real-time simulation allows developers to test and optimize navigation and control algorithms for autonomous drones in a virtual environment, accelerating development and ensuring safety.

166. What role does tactile sensing and haptic feedback play in robot control for tasks like braille writing and object recognition for visually impaired individuals?

Answer: Tactile sensing and haptic feedback enable robots to perform tasks like braille writing and object recognition for visually impaired individuals, enhancing accessibility and independence.

167. How can robots use impedance control for precise and compliant interactions in applications like rehabilitation therapy and physical assistance for the elderly?

Answer: Impedance control allows robots to provide precise and compliant interactions in applications like rehabilitation therapy and physical assistance for the elderly, improving safety and comfort.

168. What challenges and benefits arise in the control of autonomous drones for aerial surveying and mapping in remote and challenging terrain?

Answer: Autonomous drone control for aerial surveying and mapping faces challenges in navigation and endurance. Benefits include efficient data collection and mapping in remote and challenging terrains.

169. How does sensory fusion and sensor calibration enhance robot control in applications like autonomous vehicles and industrial automation?

Answer: Sensory fusion combines data from various sensors and sensor calibration ensures accuracy, enhancing robot control in applications like autonomous vehicles and industrial automation by providing reliable and precise information.

170. What is the role of reinforcement learning in robot-assisted therapy for children with autism, where personalized and adaptive interaction is crucial?
Answer: Reinforcement learning enables robots to provide personalized and adaptive interaction in robot-assisted therapy for children with autism, helping them develop social and communication skills.
171. How can robots implement hybrid force-impedance control for tasks like fine assembly and delicate object manipulation in manufacturing and electronics assembly?
Answer: Hybrid force-impedance control combines force and impedance control to perform tasks like fine assembly and delicate object manipulation with precision and sensitivity, ensuring quality in manufacturing and electronics assembly.
172. What challenges and solutions are associated with the control of aerial drones for wildlife monitoring and conservation, especially in remote and inaccessible areas?
Answer: Aerial drone control for wildlife monitoring faces challenges in navigation and battery life. Solutions include advanced navigation algorithms and efficient energy management for conservation efforts in remote and inaccessible areas.
173. How do robots employ multi-modal interaction in human-robot collaboration scenarios, enhancing communication and teamwork in industrial settings?
Answer: Multi-modal interaction enables robots to communicate through various modalities like speech, gestures, and touch, enhancing communication and teamwork in industrial settings for improved efficiency.
174. What is the impact of cognitive control in swarm robotics for tasks like disaster response and environmental monitoring, where adaptive decision-making is critical?
Answer: Cognitive control in swarm robotics enables adaptive decision-making in tasks like disaster response and environmental monitoring, improving the efficiency and effectiveness of collective actions.
175. How can robots use reinforcement learning for autonomous exploration and mapping of underground environments, such as mines and caves?
Answer: Reinforcement learning enables robots to autonomously explore and map underground environments like mines and caves, adapting to dynamic conditions and improving safety.
176. What challenges and solutions are associated with the control of swarm robots for precision agriculture, where tasks include crop monitoring and precision spraying?
Answer: Swarm robot control in precision agriculture faces challenges in communication and coordination. Solutions include swarm algorithms and distributed control for tasks like crop monitoring and precision spraying.
177. How can robots apply predictive control in robotic surgery for precise and stable movements during procedures like laparoscopic surgery and microsurgery?
Answer: Predictive control in robotic surgery uses predictive models to ensure precise and stable movements during procedures like laparoscopic surgery and microsurgery, improving surgical outcomes.
178. What is the significance of explainable AI in robot-assisted education and tutoring, where transparency and feedback are critical for effective learning?
Answer: Explainable AI in robot-assisted education provides transparent explanations and feedback to learners, enhancing their understanding and improving the effectiveness of educational robots.
179. How do robots employ real-time simulation in the development of autonomous drones for testing and optimizing navigation and control algorithms?
Answer: Real-time simulation allows developers to test and optimize navigation and control algorithms for autonomous drones in a virtual environment, accelerating development and ensuring safety.
180. What role does tactile sensing and haptic feedback play in robot control for tasks like braille writing and object recognition for visually impaired individuals?
Answer: Tactile sensing and haptic feedback enable robots to perform tasks like braille writing and object recognition for visually impaired individuals, enhancing accessibility and independence.
181. How can robots use impedance control for precise and compliant interactions in applications like rehabilitation therapy and physical assistance for the elderly?

Answer: Impedance control allows robots to provide precise and compliant interactions in applications like rehabilitation therapy and physical assistance for the elderly, improving safety and comfort.

182. What challenges and benefits arise in the control of autonomous drones for aerial surveying and mapping in remote and challenging terrain?

Answer: Autonomous drone control for aerial surveying and mapping faces challenges in navigation and endurance. Benefits include efficient data collection and mapping in remote and challenging terrains.

183. How does sensory fusion and sensor calibration enhance robot control in applications like autonomous vehicles and industrial automation?

Answer: Sensory fusion combines data from various sensors and sensor calibration ensures accuracy, enhancing robot control in applications like autonomous vehicles and industrial automation by providing reliable and precise information.

184. What is the role of reinforcement learning in robot-assisted therapy for children with autism, where personalized and adaptive interaction is crucial?

Answer: Reinforcement learning enables robots to provide personalized and adaptive interaction in robot-assisted therapy for children with autism, helping them develop social and communication skills.

185. How can robots implement hybrid force-impedance control for tasks like fine assembly and delicate object manipulation in manufacturing and electronics assembly?

Answer: Hybrid force-impedance control combines force and impedance control to perform tasks like fine assembly and delicate object manipulation with precision and sensitivity, ensuring quality in manufacturing and electronics assembly.

186. What challenges and solutions are associated with the control of aerial drones for wildlife monitoring and conservation, especially in remote and inaccessible areas?

Answer: Aerial drone control for wildlife monitoring faces challenges in navigation and battery life. Solutions include advanced navigation algorithms and efficient energy management for conservation efforts in remote and inaccessible areas.

187. How do robots employ multi-modal interaction in human-robot collaboration scenarios, enhancing communication and teamwork in industrial settings?

Answer: Multi-modal interaction enables robots to communicate through various modalities like speech, gestures, and touch, enhancing communication and teamwork in industrial settings for improved efficiency.

188. What is the impact of cognitive control in swarm robotics for tasks like disaster response and environmental monitoring, where adaptive decision-making is critical?

Answer: Cognitive control in swarm robotics enables adaptive decision-making in tasks like disaster response and environmental monitoring, improving the efficiency and effectiveness of collective actions.

189. How can robots use reinforcement learning for autonomous exploration and mapping of underground environments, such as mines and caves?

Answer: Reinforcement learning enables robots to autonomously explore and map underground environments like mines and caves, adapting to dynamic conditions and improving safety.

190. What challenges and solutions are associated with the control of swarm robots for precision agriculture, where tasks include crop monitoring and precision spraying?

Answer: Swarm robot control in precision agriculture faces challenges in communication and coordination. Solutions include swarm algorithms and distributed control for tasks like crop monitoring and precision spraying.

191. How can robots apply predictive control in robotic surgery for precise and stable movements during procedures like laparoscopic surgery and microsurgery?

Answer: Predictive control in robotic surgery uses predictive models to ensure precise and stable movements during procedures like laparoscopic surgery and microsurgery, improving surgical outcomes.

192. What is the significance of explainable AI in robot-assisted education and tutoring, where transparency and feedback are critical for effective learning?

Answer: Explainable AI in robot-assisted education provides transparent explanations and feedback to learners, enhancing their understanding and improving the effectiveness of educational robots.

193. How do robots employ real-time simulation in the development of autonomous drones for testing and optimizing navigation and control algorithms?
Answer: Real-time simulation allows developers to test and optimize navigation and control algorithms for autonomous drones in a virtual environment, accelerating development and ensuring safety.
194. What role does tactile sensing and haptic feedback play in robot control for tasks like braille writing and object recognition for visually impaired individuals?
Answer: Tactile sensing and haptic feedback enable robots to perform tasks like braille writing and object recognition for visually impaired individuals, enhancing accessibility and independence.
195. How can robots use impedance control for precise and compliant interactions in applications like rehabilitation therapy and physical assistance for the elderly?
Answer: Impedance control allows robots to provide precise and compliant interactions in applications like rehabilitation therapy and physical assistance for the elderly, improving safety and comfort.
196. What challenges and benefits arise in the control of autonomous drones for aerial surveying and mapping in remote and challenging terrain?
Answer: Autonomous drone control for aerial surveying and mapping faces challenges in navigation and endurance. Benefits include efficient data collection and mapping in remote and challenging terrains.
197. How does sensory fusion and sensor calibration enhance robot control in applications like autonomous vehicles and industrial automation?
Answer: Sensory fusion combines data from various sensors and sensor calibration ensures accuracy, enhancing robot control in applications like autonomous vehicles and industrial automation by providing reliable and precise information.
198. What is the role of reinforcement learning in robot-assisted therapy for children with autism, where personalized and adaptive interaction is crucial?
Answer: Reinforcement learning enables robots to provide personalized and adaptive interaction in robot-assisted therapy for children with autism, helping them develop social and communication skills.
199. How can robots implement hybrid force-impedance control for tasks like fine assembly and delicate object manipulation in manufacturing and electronics assembly?
Answer: Hybrid force-impedance control combines force and impedance control to perform tasks like fine assembly and delicate object manipulation with precision and sensitivity, ensuring quality in manufacturing and electronics assembly.
200. What challenges and solutions are associated with the control of aerial drones for wildlife monitoring and conservation, especially in remote and inaccessible areas?
Answer: Aerial drone control for wildlife monitoring faces challenges in navigation and battery life. Solutions include advanced navigation algorithms and efficient energy management for conservation efforts in remote and inaccessible areas.
201. How do robots utilize task planning and optimization algorithms for efficient path planning and resource allocation in applications like warehouse automation?
Answer: Task planning and optimization algorithms enable robots to efficiently plan paths and allocate resources in warehouse automation, ensuring streamlined operations and reduced resource wastage.
202. What challenges and benefits are associated with the control of robotic exoskeletons for assisting individuals with mobility impairments in daily activities?
Answer: Robotic exoskeleton control faces challenges in user adaptation and power consumption. Benefits include enhanced mobility and independence for individuals with mobility impairments.
203. How can robots apply imitation learning for skill transfer between robots and humans in collaborative tasks, such as cooking or assembly?
Answer: Imitation learning allows robots to acquire skills by observing and imitating human actions, facilitating collaboration in tasks like cooking and assembly.
204. What is the role of emotion recognition in social robots for understanding and responding to human emotions in healthcare and companion roles?

Answer: Emotion recognition equips social robots with the ability to understand and respond to human emotions, enhancing their effectiveness in healthcare and companion roles.

205. How do robots use advanced control strategies like predictive control and MPC in autonomous vehicles for safe and efficient navigation in complex urban environments?

Answer: Advanced control strategies like predictive control and Model Predictive Control (MPC) enable autonomous vehicles to navigate safely and efficiently in complex urban environments by considering dynamic factors and constraints.

206. What challenges and solutions are associated with the control of soft robots for tasks like medical procedures and exploration in unstructured environments?

Answer: Soft robot control faces challenges in modeling and actuation. Solutions include soft sensor integration and bio-inspired design for tasks like medical procedures and exploration.

207. How can robots implement fault tolerance mechanisms in critical applications like nuclear decommissioning and space exploration to ensure continued operation in the presence of failures?

Answer: Fault tolerance mechanisms enable robots to detect and recover from failures in critical applications like nuclear decommissioning and space exploration, ensuring mission success and safety.

208. What is the significance of explainable AI in military robotics for transparent decision-making and ethical considerations during combat and reconnaissance missions?

Answer: Explainable AI in military robotics provides transparent decision-making, aiding ethical considerations and accountability during combat and reconnaissance missions.

209. How do robots employ machine learning and natural language processing for human-robot interaction in customer service and information retrieval applications?

Answer: Machine learning and natural language processing enable robots to engage in human-robot interaction, facilitating customer service and information retrieval tasks.

210. What challenges and benefits arise in the control of drones for precision agriculture, including tasks like crop monitoring and pesticide application?

Answer: Drone control in precision agriculture faces challenges in data processing and regulation compliance. Benefits include enhanced crop monitoring and precise pesticide application for improved yields.

211. How can robots apply bio-inspired control in biohybrid systems for tasks like environmental monitoring and drug delivery within the human body?

Answer: Bio-inspired control in biohybrid systems enables robots to mimic biological behaviors for tasks like environmental monitoring and targeted drug delivery, increasing efficiency and adaptability.

212. What role does multi-agent coordination play in autonomous vehicles for urban transportation systems, ensuring safe and efficient traffic flow?

Answer: Multi-agent coordination in autonomous vehicles optimizes traffic flow, ensuring safe and efficient urban transportation systems by coordinating the actions of multiple vehicles.

213. How do robots utilize deep reinforcement learning in autonomous exploration of extreme environments, such as deep-sea exploration or space missions?

Answer: Deep reinforcement learning enables robots to autonomously explore extreme environments, adapting to dynamic conditions in deep-sea exploration and space missions.

214. What challenges and solutions are associated with the control of swarm robots for disaster response and search and rescue missions in complex and unpredictable environments?

Answer: Swarm robot control in disaster response faces challenges in communication and coordination. Solutions include decentralized algorithms and adaptive strategies for complex and unpredictable environments.

215. How can robots use vision-based control and computer vision techniques in autonomous drones for tasks like surveillance and environmental monitoring?

Answer: Vision-based control and computer vision techniques enable autonomous drones to perform tasks like surveillance and environmental monitoring by processing visual information for decision-making.

216. What is the impact of reinforcement learning in robot-assisted rehabilitation for stroke patients, where personalized therapy and progress tracking are essential?

Answer: Reinforcement learning enhances robot-assisted rehabilitation for stroke patients by providing personalized therapy and tracking progress, improving recovery outcomes.

217. How do robots employ hybrid locomotion mechanisms in quadrupedal robots for versatile movement in challenging terrains and environments?

Answer: Hybrid locomotion mechanisms in quadrupedal robots combine various modes of movement for versatile and adaptive mobility in challenging terrains and environments.

218. What challenges and benefits arise in the control of drones for forest fire monitoring and firefighting operations in remote and hazardous areas?

Answer: Drone control for forest fire operations faces challenges in navigation and payload capacity. Benefits include efficient monitoring and firefighting efforts in remote and hazardous areas.

219. How can robots implement swarm intelligence algorithms for collaborative exploration and mapping in unknown and cluttered environments?

Answer: Swarm intelligence algorithms enable robots to collaborate in exploration and mapping of unknown and cluttered environments, optimizing coverage and efficiency.

220. What is the role of emotion recognition in social robots for understanding and responding to human emotions in healthcare and companion roles?

Answer: Emotion recognition equips social robots with the ability to understand and respond to human emotions, enhancing their effectiveness in healthcare and companion roles.

221. How do robots use advanced control strategies like predictive control and MPC in autonomous vehicles for safe and efficient navigation in complex urban environments?

Answer: Advanced control strategies like predictive control and Model Predictive Control (MPC) enable autonomous vehicles to navigate safely and efficiently in complex urban environments by considering dynamic factors and constraints.

222. What challenges and solutions are associated with the control of soft robots for tasks like medical procedures and exploration in unstructured environments?

Answer: Soft robot control faces challenges in modeling and actuation. Solutions include soft sensor integration and bio-inspired design for tasks like medical procedures and exploration.

223. How can robots implement fault tolerance mechanisms in critical applications like nuclear decommissioning and space exploration to ensure continued operation in the presence of failures?

Answer: Fault tolerance mechanisms enable robots to detect and recover from failures in critical applications like nuclear decommissioning and space exploration, ensuring mission success and safety.

224. What is the significance of explainable AI in military robotics for transparent decision-making and ethical considerations during combat and reconnaissance missions?

Answer: Explainable AI in military robotics provides transparent decision-making, aiding ethical considerations and accountability during combat and reconnaissance missions.

225. How do robots employ machine learning and natural language processing for human-robot interaction in customer service and information retrieval applications?

Answer: Machine learning and natural language processing enable robots to engage in human-robot interaction, facilitating customer service and information retrieval tasks.

26. What challenges and benefits arise in the control of drones for precision agriculture, including tasks like crop monitoring and pesticide application?

Answer: Drone control in precision agriculture faces challenges in data processing and regulation compliance. Benefits include enhanced crop monitoring and precise pesticide application for improved yields.

227. How can robots apply bio-inspired control in biohybrid systems for tasks like environmental monitoring and drug delivery within the human body?

Answer: Bio-inspired control in biohybrid systems enables robots to mimic biological behaviors for tasks like environmental monitoring and targeted drug delivery, increasing efficiency and adaptability.

228. What role does multi-agent coordination play in autonomous vehicles for urban transportation systems, ensuring safe and efficient traffic flow?

Answer: Multi-agent coordination in autonomous vehicles optimizes traffic flow, ensuring safe and efficient urban transportation systems by coordinating the actions of multiple vehicles.

229. How do robots utilize deep reinforcement learning in autonomous exploration of extreme environments, such as deep-sea exploration or space missions?

Answer: Deep reinforcement learning enables robots to autonomously explore extreme environments, adapting to dynamic conditions in deep-sea exploration and space missions.

230. What challenges and solutions are associated with the control of swarm robots for disaster response and search and rescue missions in complex and unpredictable environments?

Answer: Swarm robot control in disaster response faces challenges in communication and coordination. Solutions include decentralized algorithms and adaptive strategies for complex and unpredictable environments.

231. How can robots use vision-based control and computer vision techniques in autonomous drones for tasks like surveillance and environmental monitoring?

Answer: Vision-based control and computer vision techniques enable autonomous drones to perform tasks like surveillance and environmental monitoring by processing visual information for decision-making.

232. What is the impact of reinforcement learning in robot-assisted rehabilitation for stroke patients, where personalized therapy and progress tracking are essential?

Answer: Reinforcement learning enhances robot-assisted rehabilitation for stroke patients by providing personalized therapy and tracking progress, improving recovery outcomes.

233. How do robots employ hybrid locomotion mechanisms in quadrupedal robots for versatile movement in challenging terrains and environments?

Answer: Hybrid locomotion mechanisms in quadrupedal robots combine various modes of movement for versatile and adaptive mobility in challenging terrains and environments.

234. What challenges and benefits arise in the control of drones for forest fire monitoring and firefighting operations in remote and hazardous areas?

Answer: Drone control for forest fire operations faces challenges in navigation and payload capacity. Benefits include efficient monitoring and firefighting efforts in remote and hazardous areas.

235. How can robots implement swarm intelligence algorithms for collaborative exploration and mapping in unknown and cluttered environments?

Answer: Swarm intelligence algorithms enable robots to collaborate in exploration and mapping of unknown and cluttered environments, optimizing coverage and efficiency.

236. How do robots use SLAM (Simultaneous Localization and Mapping) to navigate unknown environments?

Answer: SLAM enables robots to create maps of their surroundings while simultaneously determining their own position within those maps, allowing them to navigate in unfamiliar environments.

237. What types of sensors are commonly used in robot localization and mapping?

Answer: Sensors like lidar, cameras, and inertial measurement units (IMUs) are commonly used for robot localization and mapping tasks.

238. How does feature-based mapping work in robotics?

Answer: Feature-based mapping identifies and stores distinctive features in the environment, such as corners or landmarks, to create a map that the robot can use for navigation.

239. What are the advantages of grid-based mapping techniques?

Answer: Grid-based mapping divides the environment into a grid of cells, allowing for precise mapping and obstacle avoidance by the robot.

240. How does the Monte Carlo Localization (MCL) algorithm help robots estimate their position?

Answer: MCL is a probabilistic approach that uses a particle filter to estimate a robot's position by comparing sensor data with a set of particles representing possible positions.

241. What are the challenges of indoor localization in GPS-denied environments?

Answer: Indoor localization faces challenges like signal interference, multipath effects, and the lack of GPS signals, making it necessary to rely on alternative methods such as Wi-Fi, Bluetooth, or inertial sensors.

242. How do robots handle uncertainty in their localization and mapping tasks?

Answer: Robots use probabilistic methods and filters to model and mitigate uncertainty in sensor measurements and motion, ensuring accurate localization and mapping.

243. What is the role of visual SLAM in robotics, and how does it work?

Answer: Visual SLAM relies on camera images to estimate a robot's position and create a map of the environment, making it useful for robots that have cameras as primary sensors.

244. How can robots maintain accurate localization over extended periods of operation?

Answer: Robots may use loop closure detection to recognize previously visited locations, correcting accumulated localization errors and maintaining accuracy over time.

245. What is the difference between 2D and 3D mapping in robotics?

Answer: 2D mapping creates a flat representation of the environment, while 3D mapping captures depth information, enabling robots to navigate in three-dimensional spaces.

246. How does occupancy grid mapping help robots avoid obstacles?

Answer: Occupancy grid mapping divides the environment into grid cells and assigns probabilities of occupancy, helping robots identify obstacles and plan collision-free paths.

247. What are the applications of SLAM in autonomous vehicles?

Answer: SLAM is crucial for autonomous vehicles to create maps of their surroundings and accurately localize themselves for tasks like self-driving and obstacle avoidance.

248. What challenges do outdoor robots face in global localization, and how are they addressed?

Answer: Outdoor robots face challenges like GPS signal loss and varying weather conditions. Solutions may include sensor fusion and the use of additional localization techniques like visual or lidar-based SLAM.

249. How does the use of GPS improve global localization for robots?

Answer: GPS provides global coordinates that can be used to initialize a robot's position, but it may not be sufficient for precise localization, especially in urban canyons or densely populated areas.

250. How do robots handle dynamic environments when mapping and localizing?

Answer: Dynamic environments with moving objects can disrupt mapping and localization. Robots may use methods like object tracking or dynamic obstacle avoidance to adapt to changes in the environment.

251. What role does sensor fusion play in improving robot localization accuracy?

Answer: Sensor fusion combines data from multiple sensors, such as lidar, cameras, and IMUs, to improve localization accuracy and reduce the impact of individual sensor limitations.

252. How do robots perform localization and mapping in underwater environments?

Answer: Underwater robots use specialized sensors like sonar and acoustic positioning systems for localization and mapping in underwater environments.

253. How can robots handle localization and mapping in GPS-denied outdoor environments?

Answer: In GPS-denied outdoor environments, robots may use visual or lidar-based SLAM, along with sensor fusion, to localize and map their surroundings.

254. What are the limitations of landmark-based mapping techniques?

Answer: Landmark-based mapping relies on the availability of distinctive features in the environment, which may not be present in all scenarios, limiting its applicability.

255. How does topological mapping differ from geometric mapping?

Answer: Topological mapping focuses on creating a high-level representation of the environment based on connectivity and relationships between locations, while geometric mapping aims for precise geometric representations of the environment.

256. What is the role of loop closure detection in SLAM?

Answer: Loop closure detection identifies revisited locations and helps correct accumulated errors in a robot's estimated trajectory, improving the accuracy of the map.

257. How do robots handle the problem of map drift in long-duration missions?

Answer: Map drift can be addressed through loop closure detection, delocalization, and frequent map updates to maintain accuracy in long-duration missions.

258. What challenges do robots face when mapping and localizing in cluttered environments?

Answer: Cluttered environments can lead to occlusions and sensor limitations. Robots may employ advanced mapping and localization techniques to handle clutter and navigate effectively.

259. How can robots perform mapping and localization in GPS-denied underground environments?

Answer: In GPS-denied underground environments, robots may rely on inertial sensors and specialized mapping techniques to navigate and create maps.

260. What are the key factors to consider when selecting sensors for robot localization and mapping?

Answer: Factors include sensor accuracy, range, update rate, power consumption, and suitability for the environment and application.

261. How can robots adapt their mapping and localization strategies for different types of terrains, such as rough terrain or urban environments?

Answer: Robots can adapt by adjusting their sensor configurations, mapping algorithms, and navigation strategies to suit the specific challenges posed by different terrains.

262. What is the role of machine learning in improving robot localization and mapping?

Answer: Machine learning techniques can be used to enhance sensor data interpretation, optimize mapping algorithms, and improve localization accuracy in various robot applications.

263. How can robots achieve real-time mapping and localization in fast-paced environments?

Answer: Real-time mapping and localization require efficient algorithms and high-speed sensors. Robots may employ hardware acceleration or distributed computing to meet real-time requirements.

264. What challenges do robots face when localizing in GPS-denied environments with limited landmarks?

Answer: In GPS-denied environments with limited landmarks, robots may struggle to maintain accurate localization. Solutions may involve the use of additional sensors or prior map information.

265. How do robots handle situations where the environment undergoes significant changes after mapping?

Answer: Robots can adapt by detecting changes in the environment and updating their maps accordingly. Techniques like change detection algorithms help identify alterations in the surroundings.

266. What is the significance of semantic mapping in robotics?

Answer: Semantic mapping goes beyond geometric information by adding semantic labels to objects and regions in the map, providing robots with a richer understanding of their environment.

267. How can robots ensure consistent mapping and localization in environments with low light or adverse weather conditions?

Answer: Robots may employ sensor fusion, advanced imaging techniques, and additional sensors like thermal cameras to handle low-light or adverse weather conditions effectively.

268. What role does edge computing play in enhancing real-time mapping and localization for robots?

Answer: Edge computing allows robots to process sensor data locally, reducing latency and enabling real-time mapping and localization without relying heavily on remote servers.

269. How can robots handle localization challenges when operating in GPS-denied indoor environments with complex structures?

Answer: In complex indoor environments, robots can use features like visual markers, RFID tags, or indoor beacons to aid in localization and mapping.

270. What is the significance of multi-robot mapping and localization in collaborative robotics applications?

Answer: Multi-robot mapping and localization enable collaborative robots to share information and coordinate their actions, enhancing efficiency in applications like search and rescue or warehouse automation.

271. How do robots address the problem of drift in long-term autonomous missions?

Answer: To mitigate drift in long-term missions, robots employ loop closure detection techniques, frequent map updates, and delocalization. These methods help correct accumulated errors and maintain accurate localization and mapping over extended periods.

272. What challenges do robots encounter when localizing and mapping in cluttered indoor environments?

Answer: In cluttered indoor environments, robots face challenges like occlusions, limited sensor visibility, and complex geometries. They may use advanced mapping algorithms and sensor fusion to navigate effectively and create accurate maps.

273. How can robots adapt their localization and mapping strategies for underground mining operations?

Answer: Robots in underground mining adapt by using specialized sensors like laser scanners and inertial sensors. They also employ mapping techniques tailored to the unique features of underground environments.

274. What role does graph-based SLAM play in robot localization and mapping?

Answer: Graph-based SLAM models the environment as a graph, with nodes representing robot poses and edges representing measurements between poses. This approach provides a flexible framework for estimating a robot's trajectory and creating maps.

275. How can robots integrate information from external sources like GPS with their onboard sensors for global localization?

Answer: Robots integrate external information like GPS data with onboard sensors through sensor fusion techniques. This allows them to achieve more accurate global localization by combining different sources of data.

276. What are the key factors that affect the accuracy of robot localization and mapping in outdoor environments?

Answer: Key factors include GPS signal quality, sensor accuracy, terrain features, and environmental conditions. Adverse weather, signal obstructions, and sensor limitations can impact accuracy.

277. How does probabilistic localization help robots account for uncertainty in their position estimates?

Answer: Probabilistic localization assigns probability distributions to possible robot positions, enabling robots to make decisions based on the likelihood of each position. This approach accounts for uncertainty and improves accuracy.

278. What are some methods robots use to handle localization and mapping in GPS-denied underwater environments?

Answer: Robots in GPS-denied underwater environments rely on acoustic navigation, sonar mapping, and inertial sensors to navigate, localize, and map their surroundings accurately.

279. How do robots adapt their mapping and localization strategies for extreme environments like deserts or Polar Regions?

Answer: In extreme environments, robots adapt by using robust sensors and specialized algorithms for navigation. They also account for unique environmental challenges such as sand dunes or ice fields.

280. What is the role of semantic mapping in human-robot interaction and assistive robotics?

Answer: Semantic mapping helps robots understand and interact with humans by providing information about the types and functions of objects in their environment. This is valuable in assistive robotics and human-robot collaboration scenarios.

281. How do robots handle localization challenges when operating in densely populated urban environments with complex structures?

Answer: In urban environments, robots may use Wi-Fi and Bluetooth beacons, visual landmarks, and advanced SLAM techniques to navigate and localize accurately amidst complex structures and signal interference.

282. What is the significance of real-time mapping and localization in applications like autonomous vehicles and drones?

Answer: Real-time mapping and localization are critical for safe and efficient operations in applications like autonomous vehicles and drones. These systems allow vehicles to react quickly to changing environments and avoid obstacles.

283. How can robots address the limitations of sensor range and field of view in mapping and localization tasks?

Answer: Robots can employ sensor fusion, adaptive sensor configurations, and exploration strategies to overcome limitations related to sensor range and field of view, ensuring comprehensive mapping and localization.

284. What challenges and solutions are associated with multi-robot localization and mapping in large-scale environments?

Answer: Challenges include communication and coordination between robots. Solutions involve sharing map data, collaborative exploration, and managing communication networks to optimize mapping and localization.

285. How do robots handle localization and mapping when operating in confined spaces, such as tunnels or pipelines?

Answer: In confined spaces, robots use specialized sensors like 3D laser scanners and may rely on tethered or wireless communication to navigate, localize, and map accurately.

286. What techniques do robots use to adapt their localization and mapping strategies for different lighting conditions, such as low-light or bright sunlight?

Answer: Robots can adapt to varying lighting conditions by adjusting sensor settings, using additional sensors like thermal cameras, or employing advanced imaging techniques to maintain accurate mapping and localization.

287. How does sensor calibration impact the accuracy of robot localization and mapping?

Answer: Sensor calibration ensures that sensor measurements are accurate and aligned with the robot's reference frame. Proper calibration is essential for accurate mapping and localization results.

288. What challenges do robots face when mapping and localizing in environments with reflective surfaces or glass?

Answer: Reflective surfaces and glass can cause sensor interference and inaccurate readings. Robots may use materials with non-reflective properties or employ sensor filtering techniques to mitigate these challenges.

289. How can robots handle the problem of dynamic obstacles in real-time mapping and localization?

Answer: Real-time mapping and localization systems incorporate obstacle detection and tracking algorithms to identify and navigate around dynamic obstacles, ensuring safe and efficient robot operation.

290. What are the benefits of using lidar sensors in robot localization and mapping?

Answer: Lidar sensors provide accurate distance measurements and 3D point cloud data, making them valuable for precise mapping and localization in various environments, including outdoor and indoor settings.

291. How do robots perform localization and mapping when operating in environments with high electromagnetic interference, such as industrial facilities?

Answer: In high electromagnetic interference environments, robots may use shielded cables, electromagnetic interference (EMI) shielding, and noise filtering to maintain sensor accuracy and localization reliability.

292. How can robots handle localization and mapping in GPS-denied outdoor environments with dense foliage or tall buildings?

Answer: In GPS-denied outdoor environments with obstacles like dense foliage or tall buildings, robots may rely on terrain analysis, visual cues, and sensor fusion to navigate and localize accurately.

293. What is the role of predictive modeling in improving robot localization and mapping?

Answer: Predictive modeling uses historical data and sensor information to predict the robot's future position and map changes, aiding in real-time localization and mapping updates.

294. How do robots adapt their localization and mapping strategies for underwater archaeological exploration and surveying?

Answer: Robots for underwater archaeology adapt by using sonar imaging, acoustic positioning, and 3D mapping techniques tailored to the unique challenges of underwater environments.

295. What are the key considerations when choosing a mapping and localization solution for robotic applications?

Answer: Considerations include sensor capabilities, environment complexity, real-time requirements, and budget constraints. The choice depends on the specific needs of the robotic application.

296. How do robots handle localization and mapping in GPS-denied outdoor environments with complex natural features like canyons or caves?

Answer: In such environments, robots may use advanced SLAM algorithms, terrain modeling, and multi-sensor fusion to navigate and localize accurately despite the lack of GPS signals.

297. What techniques can robots use to optimize mapping and localization efficiency in large-scale outdoor environments?

Answer: Robots can optimize efficiency by employing hierarchical mapping, adaptive sampling, and distributed mapping techniques that divide the environment into manageable regions for mapping and localization.

298. How does sensor redundancy enhance the reliability of robot localization and mapping systems?

Answer: Sensor redundancy involves using multiple sensors of the same type or different types to cross-validate measurements. This redundancy enhances the reliability and robustness of mapping and localization systems.

299. What is the role of probabilistic filters like the Kalman filter in robot localization and mapping?

Answer: Probabilistic filters like the Kalman filter help estimate a robot's position and map uncertainties by continuously updating the state estimate based on sensor measurements and motion models.

300. How can robots ensure accurate localization and mapping in GPS-denied environments without relying on external infrastructure?

Answer: Robots can employ visual SLAM, inertial navigation, and terrain analysis techniques to localize and map accurately in GPS-denied environments, achieving autonomy without external infrastructure support.

301. How do robots adapt their mapping and localization strategies for underground exploration in caves or mines with intricate passages?

Answer: In caves and mines, robots use 3D mapping, lidar, and specialized sensors to navigate and localize. They also employ path planning algorithms to handle intricate passages effectively.

302. What are the challenges and solutions for robot localization and mapping in highly reflective indoor environments like glass-walled buildings?

Answer: Challenges in reflective environments include sensor interference. Solutions involve sensor filtering, using non-reflective materials, or applying anti-reflective coatings on sensors.

303. How can robots handle localization and mapping in cluttered environments with dynamic elements like moving machinery in industrial settings?

Answer: Robots use real-time obstacle detection and tracking to avoid dynamic elements in industrial settings, ensuring safe navigation and accurate mapping.

304. What is the role of semantic mapping in robotics applications related to home automation and smart environments?

Answer: Semantic mapping enables robots to recognize and interact with household objects and structures, making it valuable in home automation and smart environment applications.

305. How do robots handle localization and mapping in environments with limited visual features, such as featureless deserts or snowy plains?

Answer: In featureless environments, robots may rely on dead reckoning, inertial navigation, and terrain analysis techniques to compensate for the lack of distinct visual features.

306. What strategies do robots use to address localization challenges when operating in GPS-denied environments with tall buildings and urban canyons?

Answer: In urban canyons, robots may employ visual SLAM, Wi-Fi signal mapping, or GNSS-aided inertial navigation to handle localization challenges posed by tall buildings and signal obstructions.

307. How can robots maintain localization and mapping accuracy when navigating across different terrains, including rocky, sandy, and forested areas?

Answer: Robots adapt by adjusting their sensor configurations and fusion algorithms to account for terrain-specific challenges, ensuring accurate mapping and localization.

308. What role does sensor fusion play in optimizing robot localization and mapping performance?

Answer: Sensor fusion combines data from multiple sensors, improving localization accuracy and robustness. It helps robots navigate and map effectively in complex environments.

309. How do robots handle localization and mapping when navigating in GPS-denied underwater environments with low visibility?

Answer: In low-visibility underwater environments, robots use acoustic positioning systems, sonar imaging, and specialized sensors to localize and map accurately.

310. What are the challenges and solutions for localization and mapping in GPS-denied outdoor environments with dense foliage and vegetation?

Answer: Challenges in dense foliage include limited sensor visibility. Solutions involve terrain modeling, use of additional sensors like lidar, and sensor fusion to navigate and map accurately.

311. How can robots ensure reliable localization and mapping in underground tunnels with irregular surfaces and limited lighting?

Answer: Robots use laser scanning, stereo cameras, and illumination systems to navigate and map accurately in tunnels with irregular surfaces and limited lighting.

312. What techniques do robots employ to handle real-time mapping and localization in fast-moving autonomous vehicles and drones?

Answer: Fast-moving vehicles and drones use high-speed sensors, predictive modeling, and real-time processing to achieve responsive mapping and localization during rapid motion.

313. How does machine learning enhance the performance of robot localization and mapping algorithms?

Answer: Machine learning techniques can improve sensor data interpretation, optimize mapping algorithms, and enhance localization accuracy by learning patterns from data.

314. What role does environmental modeling play in robot localization and mapping in natural settings like forests or swamps?

Answer: Environmental modeling aids robots in predicting terrain features and adapting their navigation and mapping strategies to suit the natural setting's complexity.

315. How can robots handle localization challenges when navigating in GPS-denied indoor environments with complex layouts and tight spaces?

Answer: In indoor environments, robots use SLAM algorithms, laser range finders, and advanced sensor fusion to overcome challenges related to complex layouts and tight spaces.

316. What techniques do robots use for simultaneous localization and mapping in three-dimensional (3D) environments, such as multi-story buildings or underground structures?

Answer: In 3D environments, robots employ techniques like 3D SLAM, 3D lidar scanning, and multi-sensor fusion to perform simultaneous localization and mapping accurately.

317. How do robots address the problem of localization ambiguity when operating in environments with repetitive structures, such as warehouses with uniform aisles?

Answer: In repetitive environments, robots may employ loop closure detection and distinctive feature recognition to reduce localization ambiguity and improve mapping accuracy.

318. How can robots optimize localization and mapping strategies for applications like precision agriculture that require accurate position information for farming tasks?

Answer: In precision agriculture, robots use GNSS (Global Navigation Satellite System), specialized sensors, and terrain analysis to ensure precise localization and mapping for farming tasks.

319. What challenges and solutions are associated with real-time mapping and localization for robots in dynamic environments with rapidly changing objects?

Answer: Challenges include dynamic obstacles. Solutions involve real-time object detection, tracking, and obstacle avoidance to maintain accurate mapping and localization.

320. How do robots adapt their localization and mapping approaches for autonomous underwater vehicles (AUVs) exploring deep-sea environments?

Answer: AUVs use advanced sonar systems, acoustic positioning, and multi-beam imaging to navigate and map accurately in deep-sea environments with challenging conditions.

321. How can robots handle localization and mapping challenges when operating in GPS-denied environments with suboptimal sensor visibility, like foggy or smoky conditions?

Answer: In suboptimal visibility conditions, robots may use sensor fusion, terrain analysis, and advanced sensor filters to maintain accurate localization and mapping.

322. What are the key considerations for optimizing mapping and localization strategies for aerial drones in urban environments with complex architecture?

Answer: Considerations include obstacle detection, collision avoidance, and real-time mapping algorithms. Drones may use visual SLAM and advanced lidar sensors for accurate mapping and navigation.

323. How do robots handle the problem of localization and mapping in environments with steep slopes and rugged terrain?
Answer: In rugged terrain, robots adapt by adjusting their navigation strategies, considering elevation changes, and employing specialized sensors like inclinometers for accurate mapping and localization.
324. What is the role of virtual reality (VR) and augmented reality (AR) in improving human-robot interaction and localization assistance in complex environments?
Answer: VR and AR interfaces enhance human-robot collaboration by providing intuitive visualization of maps and robot positions in complex environments, aiding in localization assistance and navigation.
325. How can robots address localization challenges in environments with significant magnetic interference, such as industrial settings with heavy machinery?
Answer: In magnetically noisy environments, robots use magnetic field sensors and magnetic mapping techniques to compensate for interference and achieve reliable localization and mapping.
326. How do robots adapt their localization and mapping techniques for applications involving archaeological excavation and historical site preservation?
Answer: Robots for archaeological excavation employ 3D scanning, ground-penetrating radar, and non-invasive mapping techniques to preserve historical sites and artifacts while maintaining accuracy in challenging terrains.
327. What challenges and solutions are associated with robot localization and mapping in GPS-denied environments within indoor facilities with metallic structures, such as warehouses and factories?
Answer: Metallic structures can disrupt signals. Solutions involve the use of UWB (Ultra-Wideband) positioning systems, lidar, and magnetic field sensors to maintain reliable localization and mapping.
328. How can robots optimize their mapping and localization strategies for applications like infrastructure inspection of bridges, tunnels, and pipelines?
Answer: Robots use specialized sensors such as lidar and ultrasonic sensors, and may also employ rope descent techniques to access and inspect infrastructure elements accurately while mapping complex structures.
329. What techniques do robots use for localization and mapping in environments with significant temperature variations, such as desert regions or polar environments?
Answer: Robots employ thermal imaging, temperature sensors, and adaptive navigation strategies to compensate for temperature variations and maintain accurate localization and mapping.
330. What role does multi-modal sensor fusion play in improving the robustness of robot localization and mapping systems?
Answer: Multi-modal sensor fusion combines data from different types of sensors, such as visual, inertial, and lidar, to enhance localization accuracy and resilience against sensor failures.
331. How do robots address localization challenges when operating in cluttered environments with narrow passages, such as underground pipelines or ventilation ducts?
Answer: Robots use miniature sensors, obstacle detection, and path planning algorithms to navigate and map accurately in narrow and cluttered passages.
332. What are the benefits of using passive sensors like passive RFID tags for indoor robot localization and mapping?
Answer: Passive RFID tags provide cost-effective and energy-efficient solutions for indoor robot localization and mapping, allowing robots to detect and recognize tagged landmarks.
333. How can robots handle localization and mapping challenges in environments with limited or unstable wireless connectivity, such as remote wilderness areas or disaster-stricken regions?
Answer: Robots use terrain-based navigation, satellite communication, and adaptive routing strategies to overcome challenges related to limited or unstable wireless connectivity in such environments.
334. What techniques do robots employ for real-time mapping and localization in urban transportation systems, including subway networks and bus terminals?
Answer: Robots use Wi-Fi signal mapping, inertial navigation, and sensor fusion to navigate and localize accurately within complex urban transportation systems.

335. How do robots adapt their localization and mapping strategies for precision medical procedures, such as minimally invasive surgery and targeted drug delivery?

Answer: Robots in precision medical procedures use specialized sensors and medical imaging techniques for precise localization and mapping, ensuring accuracy in critical applications.

336. What challenges and solutions are associated with multi-floor localization and mapping in buildings with multiple stories?

Answer: Challenges include floor transitions. Solutions involve recognizing elevators, stairs, and detecting changes in floor levels to maintain accurate multi-floor localization and mapping.

337. How can robots ensure localization and mapping accuracy when operating in environments with changing magnetic fields, such as near electrical machinery or power lines?

Answer: Robots use magnetometer sensors and magnetic field mapping techniques to adapt to changing magnetic fields and maintain accurate localization and mapping.

338. What role does cloud-based mapping and localization play in enabling collaborative robotic systems to share map data and coordinate their actions?

Answer: Cloud-based mapping and localization systems enable collaborative robots to share map data in real-time, enhancing coordination and efficiency in various applications.

339. How do robots handle localization challenges in environments with unstable terrain, such as sandy beaches or marshlands?

Answer: Robots adapt by employing techniques like wheel or leg designs optimized for unstable terrain and may use additional sensors like pressure sensors to navigate and map effectively.

340. What is the significance of map annotation in robotics, and how does it enhance human-robot interaction?

Answer: Map annotation allows robots to add information, such as labels or notes, to their maps, facilitating better communication with humans and improving the effectiveness of human-robot interaction.

341. How can robots maintain accurate localization and mapping in environments with high levels of radiofrequency interference, such as radio towers or communication hubs?

Answer: Robots may employ frequency-hopping radios, signal filtering, and adaptive communication protocols to mitigate the impact of radiofrequency interference on localization and mapping.

342. What are the challenges and solutions for localization and mapping in environments with significant acoustic interference, such as noisy industrial facilities or crowded concert venues?

Answer: Challenges include noise interference. Solutions involve using acoustic sensors with noise-cancellation capabilities and advanced signal processing techniques to achieve reliable localization and mapping.

343. How do robots adapt their localization and mapping strategies for applications in the field of precision agriculture to optimize crop management and yield?

Answer: Robots use GNSS, soil sensors, and crop monitoring systems to achieve precise localization and mapping, enabling targeted crop management and increased yield.

344. What role does machine learning play in improving the efficiency and accuracy of robot localization and mapping in complex environments?

Answer: Machine learning is used for sensor data interpretation, obstacle recognition, and map updates, enhancing efficiency and accuracy in robot localization and mapping tasks.

345. How can robots ensure safe navigation and accurate mapping in environments with low or zero visibility, such as smoke-filled buildings during firefighting operations?

Answer: Robots use thermal imaging, gas sensors, and communication systems to navigate and map accurately in low-visibility environments, ensuring safety during firefighting and rescue missions.

346. What techniques do robots employ to handle localization and mapping challenges in environments with highly reflective or transparent materials, such as glass and mirrors?

Answer: Robots use materials with non-reflective properties, such as anti-reflective coatings, and may employ advanced sensor filtering techniques to mitigate challenges posed by highly reflective or transparent materials.

347. How can robots optimize their localization and mapping strategies for agricultural robotics applications, including autonomous tractors and harvesters?
Answer: In agricultural robotics, robots use GPS-guided navigation, soil sensors, and crop monitoring systems to optimize localization and mapping for precision farming tasks.
348. What is the role of probabilistic graphical models in enhancing robot localization and mapping accuracy?
Answer: Probabilistic graphical models, such as Markov random fields, enable robots to incorporate spatial relationships and dependencies between map elements, improving accuracy in localization and mapping.
349. How do robots handle localization and mapping challenges when operating in confined spaces with complex geometries, such as underground pipelines or ventilation ducts?
Answer: Robots use 3D mapping, lidar, and adaptive navigation algorithms to navigate and map accurately in confined spaces with intricate geometries.
350. What are the key considerations for adapting robot localization and mapping strategies to accommodate changing environmental conditions, such as seasonal variations or weather fluctuations?
Answer: Considerations include sensor calibration, weather monitoring, and predictive modeling. Robots may use adaptive algorithms to handle changes in environmental conditions effectively.
351. How can robots enhance localization and mapping accuracy in environments with potential signal interference from electronic devices or power lines?
Answer: Robots employ electromagnetic interference (EMI) shielding and signal filtering techniques to mitigate interference effects, ensuring accurate localization and mapping in such environments.
352. What challenges and solutions are associated with localization and mapping for robot applications in extreme environments, including deep-sea exploration and space missions?
Answer: Challenges include extreme pressure and temperature conditions. Solutions involve using specialized sensors and materials suitable for extreme environments to maintain accurate localization and mapping.
353. How do robots handle localization challenges when operating in environments with rapidly changing lighting conditions, such as indoor facilities with large windows and skylights?
Answer: Robots use adaptive exposure control for cameras and may employ additional sensors like thermal cameras to handle rapid lighting changes and maintain accurate localization and mapping.
354. What is the significance of collaborative mapping and localization for multi-robot systems in applications like swarm robotics and search and rescue missions?
Answer: Collaborative mapping and localization enable multi-robot systems to share maps and coordinate their efforts, improving efficiency and coverage in tasks like search and rescue.
355. How can robots adapt their localization and mapping strategies for applications in precision construction and infrastructure projects?
Answer: Robots use advanced sensors like 3D lidar and GNSS, as well as building information modeling (BIM) data, to achieve precise localization and mapping in construction and infrastructure projects.
356. What techniques do robots employ to handle localization and mapping challenges when navigating through heavily wooded areas or dense forests?
Answer: Robots use terrain analysis, lidar sensors, and adaptive planning algorithms to navigate and map accurately in heavily wooded areas and dense forests.
357. What role does semantic segmentation play in enhancing robot perception and localization in urban environments with complex signage and road markings?
Answer: Semantic segmentation allows robots to recognize and interpret road markings and signage, aiding in localization and navigation in complex urban environments.
358. How can robots optimize localization and mapping strategies for underwater applications, including marine research and underwater archaeology?
Answer: Underwater robots use acoustic positioning, sonar imaging, and bathymetric mapping techniques to optimize localization and mapping for marine research and underwater archaeology.

359. What are the challenges and solutions for localization and mapping in environments with hazardous materials, such as chemical plants or nuclear facilities?

Answer: Challenges include safety considerations. Solutions involve remote operation, specialized sensors, and protective measures to ensure safe localization and mapping in hazardous environments.

360. How do robots address localization challenges when navigating through underground sewage systems, pipelines, or utility tunnels with complex networks?

Answer: Robots use 3D mapping, inertial navigation, and multi-sensor fusion to navigate and map accurately within complex underground networks and confined spaces.

361. What is the significance of localization and mapping accuracy for autonomous aerial vehicles (drones) in applications like aerial photography and surveying?

Answer: Localization and mapping accuracy is critical for drones in capturing precise aerial imagery and conducting accurate surveys for applications in agriculture, forestry, and land mapping.

362. How can robots adapt their localization and mapping approaches for applications in cultural heritage preservation, including the exploration of historical sites and artworks?

Answer: Robots use non-invasive sensing technologies, such as 3D laser scanning and multispectral imaging, to preserve cultural heritage sites and artifacts while mapping them accurately.

363. What techniques do robots employ to handle real-time mapping and localization in dynamic environments like theme parks and entertainment venues?

Answer: Robots use dynamic obstacle detection, path planning, and real-time map updates to navigate and localize accurately in dynamic entertainment venues.

364. How do robots handle localization challenges when operating in GPS-denied indoor environments with intricate machinery and conveyor systems, such as factories and warehouses?

Answer: Robots use indoor positioning systems, lidar, and sensor fusion to maintain accurate localization and mapping in indoor environments with complex machinery.

365. What are the key considerations for optimizing robot localization and mapping strategies in applications involving autonomous underwater exploration, such as marine biology research?

Answer: Considerations include acoustic communication, sonar mapping, and adaptive navigation algorithms, enabling robots to excel in autonomous underwater exploration tasks for marine biology research.

366. How can robots ensure accurate localization and mapping when operating in environments with unpredictable terrain, such as post-disaster recovery sites or wilderness areas?

Answer: Robots employ robust path planning, adaptive sensors, and terrain analysis techniques to adapt to unpredictable terrain and maintain accurate localization and mapping.

367. What is the role of cloud-based data sharing platforms in collaborative mapping and localization for robot fleets in logistics and delivery applications?

Answer: Cloud-based platforms enable real-time data sharing among robot fleets, enhancing collaboration and optimization in logistics and delivery operations.

368. How do robots handle localization challenges in environments with complex geometry, such as historical buildings with intricate architecture and irregular layouts?

Answer: Robots use 3D mapping, lidar, and visual SLAM techniques to navigate and map accurately within historical buildings with complex geometry.

369. What challenges and solutions are associated with robot localization and mapping in environments with high levels of electromagnetic interference, such as radio frequency (RF) shielding facilities?

Answer: Challenges include signal distortion. Solutions involve using sensors with high immunity to interference and adapting localization algorithms to account for signal distortion.

370. How can robots optimize their localization and mapping strategies for applications in precision mining and mineral exploration?

Answer: Robots use geophysical sensors, satellite positioning, and geological data to optimize localization and mapping for precision mining and mineral exploration tasks.

371. What role does augmented reality (AR) play in enhancing robot localization and mapping for tasks involving remote assistance and teleoperation?

Answer: AR interfaces provide real-time visualization of robot positions and maps, aiding in remote assistance and teleoperation tasks by enhancing situational awareness.

372. How do robots address localization challenges in environments with changing lighting conditions caused by natural phenomena like sun glare or shadowing?

Answer: Robots use adaptive sensor settings and advanced lighting compensation techniques to handle changing lighting conditions caused by natural phenomena and maintain accurate localization and mapping.

373. What techniques do robots employ for real-time mapping and localization in environments with intricate sewer systems and underground drainage networks?

Answer: Robots use 3D mapping, sewer inspection cameras, and adaptive navigation to operate efficiently and map accurately within intricate sewer systems and underground networks.

374. How can robots ensure accurate localization and mapping in environments with rapidly varying water currents, such as riverbed surveys and underwater monitoring?

Answer: Robots use underwater current sensors, advanced sonar systems, and adaptive navigation strategies to adapt to rapidly varying water currents and maintain precise localization and mapping.

375. What is the significance of map consistency and loop closure detection in improving robot localization and mapping accuracy?

Answer: Map consistency and loop closure detection techniques help robots identify and correct errors in their maps, leading to enhanced localization and mapping accuracy.

376. How do robots adapt their localization and mapping strategies for applications in the agricultural industry, including precision planting and harvesting?

Answer: Robots in agriculture use GPS-guided navigation, soil sensors, and crop monitoring systems to optimize localization and mapping for precision planting and harvesting tasks.

377. What challenges and solutions are associated with localization and mapping in environments with fluctuating magnetic anomalies, such as regions with volcanic activity?

Answer: Challenges include magnetic anomalies. Solutions involve sensor calibration, anomaly detection, and mapping techniques designed to handle fluctuating magnetic fields.

378. How can robots handle localization challenges when navigating through intricate mazes or labyrinthine structures, such as hedge mazes in botanical gardens?

Answer: Robots employ path planning algorithms, exploration strategies, and obstacle detection to navigate accurately through intricate mazes and labyrinthine structures.

379. What are the key considerations for adapting robot localization and mapping strategies to accommodate seasonal changes in vegetation and landscape, such as agricultural fields?

Answer: Considerations include vegetation modeling, seasonal mapping updates, and adaptive navigation algorithms, enabling robots to adapt to changing landscapes and maintain accurate localization and mapping.

380. How can robots optimize their localization and mapping approaches for tasks in archaeological excavation, including the exploration of buried artifacts and historical sites?

Answer: Robots use ground-penetrating radar, 3D scanning, and archaeological data to optimize localization and mapping for excavation and historical site exploration, preserving cultural heritage with precision.

381. How do robots address localization challenges when operating in highly reflective underwater environments, such as coral reefs or submerged metal structures?

Answer: Robots utilize underwater-friendly sensors and advanced computer vision techniques to mitigate reflection issues, ensuring precise localization and mapping in reflective underwater environments.

382. What challenges and solutions are associated with localization and mapping for swarm robotics applications, where multiple robots collaborate to achieve common goals?

Answer: Challenges include communication and coordination. Solutions involve swarm algorithms, shared maps, and distributed localization strategies to ensure effective collaboration and mapping in swarm robotics.

383. How can robots optimize their localization and mapping strategies for applications in autonomous mining, including ore extraction and mineral resource management?

Answer: Robots use geospatial sensors, ore sensors, and geological data to optimize localization and mapping for autonomous mining operations, enhancing resource management and extraction efficiency.

384. How do robots handle localization challenges when navigating in environments with highly dynamic elements, such as crowded shopping malls or busy airports?

Answer: Robots use real-time object tracking, crowd analysis, and navigation prediction algorithms to safely navigate and accurately map crowded environments with dynamic elements.

385. What is the role of human-robot interaction (HRI) in improving robot localization and mapping accuracy in collaborative scenarios, like warehouse automation?

Answer: HRI enables humans to provide localization cues and assist robots in ambiguous situations, enhancing localization and mapping accuracy in collaborative warehouse automation.

386. How can robots ensure accurate localization and mapping in environments with extensive underground networks, such as subways and underground utility systems?

Answer: Robots utilize subway maps, inertial navigation systems, and underground mapping techniques to accurately localize and map within extensive underground networks.

387. What techniques do robots employ for real-time mapping and localization in dynamic urban environments with heavy traffic and pedestrians?

Answer: Robots use traffic analysis, pedestrian detection, and real-time map updates to navigate safely and localize accurately in dynamic urban environments with heavy traffic.

388. How do robots address localization challenges when operating in environments with magnetic anomalies caused by geological features, such as mineral deposits?

Answer: Robots use magnetic anomaly detection and compensation techniques, combined with advanced navigation algorithms, to handle localization challenges in areas with magnetic anomalies.

389. What are the key considerations for optimizing robot localization and mapping strategies for applications in precision forestry, including tree monitoring and logging?

Answer: Considerations include lidar-based tree scanning, GPS-guided navigation, and real-time mapping updates, enabling robots to excel in precision forestry tasks.

390. How can robots ensure accurate localization and mapping when operating in GPS-denied underwater environments with strong currents and turbulent water conditions?

Answer: Robots use underwater current sensors, acoustic positioning, and advanced control systems to adapt to strong currents and maintain precise localization and mapping.

391. What is the significance of sensor calibration and drift correction in improving the long-term accuracy of robot localization and mapping systems?

Answer: Sensor calibration and drift correction techniques help robots maintain accuracy over time, reducing errors and enhancing the long-term reliability of localization and mapping systems.

392. How do robots adapt their localization and mapping strategies for applications in the field of marine biology, including the study of underwater ecosystems?

Answer: Robots use underwater cameras, sonar imaging, and environmental sensors to optimize localization and mapping for marine biology research, aiding in the study of underwater ecosystems.

393. What challenges and solutions are associated with robot localization and mapping in environments with highly reflective surfaces, such as mirrored skyscrapers or polished floors?

Answer: Challenges include sensor interference. Solutions involve sensor filtering, adaptive algorithms, and reflective surface recognition to handle localization and mapping in such environments.

394. How can robots handle localization challenges when operating in GPS-denied underground mining tunnels with irregular shapes and layouts?

Answer: Robots use underground mapping systems, laser scanners, and adaptive navigation strategies to accurately localize and map within irregular underground mining tunnels.

395. What role does human pose estimation play in improving the accuracy of robot localization and mapping in human-robot interaction scenarios?

Answer: Human pose estimation helps robots understand the position and movement of humans, enhancing accuracy and safety in human-robot interaction contexts.

396. How do robots address localization challenges when navigating in environments with rapidly changing weather conditions, such as outdoor construction sites?

Answer: Robots use weather monitoring sensors, adaptive path planning, and real-time updates to handle localization challenges caused by rapidly changing weather conditions.

397. What techniques do robots employ for real-time mapping and localization in industrial settings with complex machinery, such as manufacturing plants and refineries?

Answer: Robots use 3D lidar, machine vision systems, and machine interface integration to navigate safely and localize accurately in industrial settings with complex machinery.

398. How can robots ensure accurate localization and mapping in environments with high levels of electromagnetic interference from electronic devices, such as hospitals and research labs?

Answer: Robots utilize EMI-resistant sensors and interference-tolerant algorithms to mitigate the effects of electromagnetic interference, ensuring precise localization and mapping in such environments.

399. What are the challenges and solutions for localization and mapping in environments with rapidly changing lighting conditions due to artificial sources, such as discotheques or concert halls?

Answer: Challenges include extreme lighting changes. Solutions involve adaptive sensor settings, dynamic exposure control, and real-time lighting compensation techniques for accurate localization and mapping.

400. How do robots adapt their localization and mapping strategies for applications in the field of marine archaeology, including the exploration of shipwrecks and underwater relics?

Answer: Robots use underwater archaeology sensors, 3D scanning, and sonar imaging to optimize localization and mapping for marine archaeology, preserving historical artifacts with precision.

401. What is the significance of probabilistic modeling in robot localization and mapping, and how does it contribute to robustness in uncertain environments?

Answer: Probabilistic models allow robots to represent uncertainty, improving localization and mapping accuracy in uncertain or dynamic environments by making informed probabilistic estimates.

402. How can robots address localization challenges when navigating through intricate underground sewer systems with narrow passages and branching tunnels?

Answer: Robots use miniature sensors, advanced mapping techniques, and adaptive path planning to navigate accurately in narrow passages and complex sewer systems.

403. What challenges and solutions are associated with robot localization and mapping in environments with heavy vegetation and dense foliage, such as rainforests or jungles?

Answer: Challenges include limited sensor visibility. Solutions involve terrain analysis, lidar sensors, and multi-sensor fusion to navigate and map accurately in dense foliage.

404. How do robots optimize their localization and mapping strategies for applications in the oil and gas industry, including pipeline inspection and maintenance?

Answer: Robots use specialized sensors like pipeline inspection tools, GPS guidance, and catholic protection systems to optimize localization and mapping for oil and gas industry tasks.

405. What role does simultaneous localization and mapping (SLAM) play in improving robot navigation and mapping in real-time scenarios?

Answer: SLAM techniques enable robots to simultaneously estimate their position and build a map of their surroundings, enhancing navigation and mapping in real-time applications.

406. How can robots ensure accurate localization and mapping when operating in environments with severe temperature fluctuations, such as deserts or Polar Regions?

Answer: Robots use temperature-resistant sensors, thermal imaging, and adaptive algorithms to compensate for temperature fluctuations and maintain precise localization and mapping.

407. What are the key considerations for adapting robot localization and mapping strategies to accommodate seasonal variations in lighting conditions, such as those encountered in outdoor agriculture?

Answer: Considerations include seasonal lighting models, adaptive sensors, and real-time lighting compensation, allowing robots to adapt to seasonal variations and maintain accurate localization and mapping.

408. How do robots address localization challenges when navigating through complex, multi-level parking structures with limited GPS signal availability?

Answer: Robots use indoor positioning systems, lidar sensors, and multi-level mapping techniques to navigate accurately within multi-level parking structures with limited GPS signals.

409. What is the role of computer vision in enhancing robot localization and mapping in applications involving object recognition and manipulation?

Answer: Computer vision enables robots to recognize objects, improving object manipulation accuracy and contributing to better localization and mapping in tasks requiring object interaction.

410. How can robots optimize their localization and mapping approaches for applications in precision viticulture, including vineyard management and grape harvesting?

Answer: Robots use GPS-based vineyard mapping, soil sensors, and grape ripeness sensors to optimize localization and mapping for precision viticulture, enhancing vineyard management and grape harvesting efficiency.

411. What challenges and solutions are associated with robot localization and mapping in cluttered home environments with various furniture and objects?

Answer: Challenges include obstacle avoidance. Solutions involve using depth cameras, lidar sensors, and machine learning algorithms to navigate safely and map accurately in cluttered home environments.

412. How do robots handle localization challenges when operating in environments with frequent electromagnetic interference from household appliances and electronic devices?

Answer: Robots employ interference-resistant sensors and adaptive signal processing to mitigate electromagnetic interference effects, ensuring precise localization and mapping in household environments.

413. What techniques do robots employ for real-time mapping and localization in large-scale outdoor events, such as music festivals or sporting events?

Answer: Robots use crowd analysis, event-specific maps, and real-time updates to navigate safely and localize accurately in dynamic and crowded outdoor event venues.

414. How can robots ensure accurate localization and mapping in environments with rapid changes in lighting conditions due to artificial sources, such as photography studios or film sets?

Answer: Robots utilize real-time lighting analysis, adaptive camera settings, and lighting correction techniques to adapt to rapid lighting changes and maintain accurate localization and mapping.

415. What is the significance of visual odometer in improving robot localization and mapping accuracy, particularly in scenarios with limited sensor range?

Answer: Visual odometer estimates robot motion by tracking visual features, aiding in localization and mapping accuracy in scenarios where sensor range may be limited or intermittent.

416. How do robots adapt their localization and mapping strategies for applications in environmental monitoring, including the study of ecosystems and wildlife habitats?

Answer: Robots use environmental sensors, wildlife monitoring tools, and GIS data to optimize localization and mapping for environmental monitoring, contributing to ecosystem and habitat research.

417. What challenges and solutions are associated with localization and mapping in environments with highly variable terrain, such as mountainous regions or rocky landscapes?

Answer: Challenges include unstable terrain. Solutions involve terrain analysis, adaptive suspension systems, and multi-sensor fusion to handle localization and mapping in variable terrains.

418. How can robots handle localization challenges when navigating in environments with dense smoke or fog, such as during search and rescue operations in burning buildings?

Answer: Robots use thermal imaging, gas sensors, and adaptive navigation strategies to navigate and map accurately in low-visibility environments with dense smoke or fog.

419. What role does semantic mapping play in enhancing robot perception and localization in environments with complex structures and objects, such as warehouses?
Answer: Semantic mapping allows robots to recognize and categorize objects and structures, aiding in perception and localization in complex environments like warehouses.
420. How do robots address localization challenges when operating in environments with rapidly changing atmospheric conditions, such as wind patterns in outdoor agricultural fields?
Answer: Robots use weather monitoring sensors, wind compensation algorithms, and adaptive navigation strategies to handle localization challenges caused by rapidly changing atmospheric conditions.
421. What techniques do robots employ for real-time mapping and localization in construction sites with dynamically changing layouts and structures?
Answer: Robots use construction site monitoring, 3D lidar scanning, and building information modeling (BIM) integration to navigate safely and localize accurately in dynamic construction environments.
422. How can robots ensure accurate localization and mapping in environments with frequent signal interference from nearby electronic devices, such as in busy urban areas?
Answer: Robots employ signal filtering, interference detection, and adaptive communication protocols to mitigate signal interference effects and maintain precise localization and mapping.
423. What are the key considerations for optimizing robot localization and mapping strategies in applications involving precision inspection and quality control in manufacturing?
Answer: Considerations include sensor integration, inspection algorithms, and real-time quality control feedback, enabling robots to excel in precision inspection tasks within manufacturing environments.
424. How do robots adapt their localization and mapping strategies for applications in the agricultural industry, such as autonomous fruit picking and orchard management?
Answer: Robots in agriculture use GPS-guided navigation, fruit detection sensors, and orchard mapping to optimize localization and mapping for tasks like autonomous fruit picking and orchard management.
425. What challenges and solutions are associated with robot localization and mapping in underwater environments with limited visibility, such as murky lakes or reservoirs?
Answer: Challenges include limited visibility. Solutions involve sonar-based navigation, underwater mapping techniques, and adaptive sensors to handle localization and mapping in such environments.
426. What is the role of deep learning in improving the recognition of complex objects and landmarks during robot localization and mapping tasks?
Answer: Deep learning enables robots to recognize and interpret complex objects and landmarks, enhancing object-based localization and mapping accuracy in various scenarios.
427. How can robots address localization challenges when operating in dynamic agricultural fields with constantly changing crop growth patterns and conditions?
Answer: Robots use real-time crop monitoring, adaptive path planning, and dynamic mapping updates to handle localization challenges in dynamic agricultural fields with changing crop conditions.
428. What challenges and solutions are associated with robot localization and mapping in environments with significant acoustic interference from machinery or equipment noise?
Answer: Challenges include noise interference. Solutions involve using acoustic sensors with noise-cancellation capabilities and advanced signal processing techniques to achieve reliable localization and mapping.

429. How do robots optimize their localization and mapping approaches for tasks in marine aquaculture, including fish farming and underwater habitat monitoring?
Answer: Robots use underwater aquaculture sensors, underwater cameras, and hydro acoustic mapping to optimize localization and mapping for marine aquaculture operations, supporting fish farming and habitat monitoring.
430. What is the significance of adaptive sensor fusion in enhancing robot localization and mapping accuracy, especially in scenarios with diverse sensor types?
Answer: Adaptive sensor fusion allows robots to intelligently combine data from different sensor types, improving localization and mapping accuracy in complex and diverse sensor environments.

431. How can robots ensure accurate localization and mapping in environments with challenging topography, such as steep cliffs or rocky coastlines?

Answer: Robots use terrain analysis, cliff detection algorithms, and lidar sensors to navigate safely and map accurately in challenging topographic environments.

432. What techniques do robots employ for real-time mapping and localization in archaeological excavations, including the exploration of ancient ruins and historical artifacts?

Answer: Robots use 3D laser scanning, archaeological data integration, and artifact recognition to navigate and map accurately in archaeological excavations, preserving historical artifacts and sites.

433. How do robots handle localization challenges when operating in environments with rapidly changing water levels, such as river flood zones or coastal areas prone to tides?

Answer: Robots use water level sensors, tidal predictions, and adaptive navigation algorithms to adapt to changing water levels and maintain accurate localization and mapping.

434. What role does augmented reality (AR) play in enhancing robot localization and mapping for tasks involving visual cues and annotations?

Answer: AR overlays visual cues and annotations on robot maps, aiding in localization and navigation by providing contextual information in real-time.

435. How can robots optimize their localization and mapping strategies for applications in precision horticulture, including greenhouse automation and plant monitoring?

Answer: Robots use greenhouse-specific sensors, plant health monitoring tools, and precision mapping to optimize localization and mapping for precision horticulture tasks, improving plant care and crop yield.

436. What challenges and solutions are associated with robot localization and mapping in environments with highly reflective surfaces, such as glass facades in urban areas?

Answer: Challenges include sensor reflection. Solutions involve sensor angle adjustments, polarization filters, and advanced algorithms to mitigate reflection effects and maintain accurate localization and mapping.

437. How do robots handle localization challenges when navigating through confined spaces with limited clearance, such as narrow tunnels or pipeline inspections?

Answer: Robots use miniaturized sensors, obstacle detection algorithms, and adaptive motion planning to navigate accurately in confined spaces with limited clearance.

438. What techniques do robots employ for real-time mapping and localization in agricultural greenhouses with complex plant arrangements and varying lighting conditions?

Answer: Robots use plant recognition algorithms, lighting control systems, and adaptive mapping to navigate and map accurately in dynamic agricultural greenhouse environments.

439. How can robots ensure accurate localization and mapping in environments with frequent variations in temperature and humidity, such as tropical rainforests?

Answer: Robots use humidity and temperature sensors, environmental modeling, and adaptive navigation to compensate for variations in temperature and humidity, maintaining precise localization and mapping.

440. What is the significance of multi-modal sensor fusion in improving robot localization and mapping accuracy in environments with diverse sensor types?

Answer: Multi-modal sensor fusion enables robots to combine data from various sensors, enhancing localization and mapping accuracy in complex environments with diverse sensor types.

441. How do robots adapt their localization and mapping strategies for applications in archaeological conservation, including the preservation of fragile artifacts and historical sites?

Answer: Robots use non-invasive sensors, artifact recognition, and conservation guidelines to optimize localization and mapping for archaeological conservation, aiding in artifact preservation.

442. What challenges and solutions are associated with robot localization and mapping in environments with rapidly changing atmospheric pressure, such as deep-sea exploration?

Answer: Challenges include pressure variation. Solutions involve pressure-resistant sensors, adaptive control systems, and deep-sea mapping techniques to handle localization and mapping in high-pressure environments.

443. How can robots address localization challenges when operating in environments with irregular or shifting magnetic fields, such as regions with geological anomalies?
Answer: Robots use magnetic anomaly modeling, field mapping, and adaptive localization algorithms to adapt to irregular or shifting magnetic fields and maintain accurate localization and mapping.
444. What role does predictive modeling play in improving robot navigation and mapping in dynamic environments with moving objects, such as busy city streets?
Answer: Predictive modeling allows robots to anticipate the movement of objects, enhancing navigation and mapping in dynamic environments with moving obstacles.
445. How do robots handle localization challenges when operating in environments with rapidly changing water salinity levels, such as estuaries and coastal areas?
Answer: Robots use salinity sensors, water quality monitoring, and adaptive navigation to adapt to changing water salinity levels and maintain accurate localization and mapping.
446. What techniques do robots employ for real-time mapping and localization in the mining industry, including underground tunnel mapping and ore extraction?
Answer: Robots use underground mapping systems, ore detection sensors, and 3D lidar scanning to navigate and map accurately in mining environments, supporting tunnel mapping and ore extraction.
447. How can robots ensure accurate localization and mapping in environments with extreme temperature differentials between day and night, such as desert regions?
Answer: Robots use thermal management systems, temperature modeling, and adaptive algorithms to compensate for temperature differentials and maintain precise localization and mapping.
448. What are the key considerations for optimizing robot localization and mapping strategies in precision agriculture applications, such as soil analysis and crop spraying?
Answer: Considerations include soil sensing, variable rate technology, and real-time crop monitoring, enabling robots to excel in precision agriculture tasks like soil analysis and crop spraying.
449. How do robots address localization challenges when navigating in environments with rapidly changing wind patterns, such as wind farms or coastal areas?
Answer: Robots use wind monitoring sensors, wind compensation algorithms, and adaptive navigation strategies to handle localization challenges caused by rapidly changing wind patterns.
450. What is the significance of 3D point cloud mapping in enhancing robot localization and mapping accuracy, particularly in environments with complex structures?
Answer: 3D point cloud mapping enables robots to create detailed and accurate maps of complex structures, enhancing localization and mapping accuracy in environments with intricate geometry.
451. How can robots optimize their localization and mapping approaches for applications in the field of marine geology, including seafloor mapping and geological surveys?
Answer: Robots use seafloor mapping sensors, geological sampling tools, and adaptive navigation to optimize localization and mapping for marine geology research, aiding in seafloor mapping and geological surveys.
452. What challenges and solutions are associated with robot localization and mapping in environments with frequent signal interference from neighboring robots or wireless devices?
Answer: Challenges include signal collisions. Solutions involve communication protocols, interference detection, and adaptive localization algorithms to handle interference effects and maintain accurate localization and mapping.
453. How do robots handle localization challenges when operating in environments with rapidly changing water turbidity, such as riverbed surveys in sediment-laden waters?
Answer: Robots use turbidity sensors, adaptive vision systems, and real-time mapping updates to adapt to changing water turbidity levels and maintain accurate localization and mapping.
454. What role does natural language processing (NLP) play in enhancing robot-human communication during localization and mapping tasks in collaborative settings?
Answer: NLP enables robots to understand and respond to human commands and queries, enhancing communication and cooperation during localization and mapping in collaborative environments.

455. How can robots ensure accurate localization and mapping in environments with frequent vibration and mechanical disturbances, such as construction sites with heavy machinery?

Answer: Robots use vibration-resistant sensors, motion compensation algorithms, and adaptive navigation to handle localization challenges caused by frequent vibrations and mechanical disturbances.

456. What techniques do robots employ for real-time mapping and localization in precision healthcare settings, such as surgery and medical device navigation?

Answer: Robots use specialized medical sensors, image-guided navigation, and real-time updates to navigate safely and localize accurately in precision healthcare environments, supporting surgical procedures and medical device navigation.

457. How do robots address localization challenges when operating in environments with rapidly changing geological conditions, such as volcanic terrain with lava flows?

Answer: Robots use geological monitoring, lava flow detection sensors, and adaptive navigation to adapt to changing geological conditions and maintain accurate localization and mapping.

458. What is the significance of cloud-based map storage and retrieval in improving the scalability and efficiency of robot localization and mapping systems?

Answer: Cloud-based storage allows robots to access and share maps, enhancing scalability and collaboration in large-scale localization and mapping systems.

459. How can robots optimize their localization and mapping approaches for applications in precision forestry, including timber harvesting and forest management?

Answer: Robots use GPS-based forest mapping, tree health sensors, and real-time forest monitoring to optimize localization and mapping for precision forestry tasks, supporting timber harvesting and forest management.

460. How do robots handle localization challenges when navigating through environments with rapidly changing water flow patterns, such as river navigation in changing river currents?

Answer: Robots use water flow sensors, hydrodynamic modeling, and adaptive navigation to adapt to changing water flow patterns and maintain accurate localization and mapping.

461. What challenges and solutions are associated with robot localization and mapping in environments with complex acoustics, such as concert halls or auditoriums?

Answer: Challenges include sound reflections. Solutions involve acoustic modeling, sound source localization, and adaptive algorithms to mitigate acoustic effects and maintain accurate localization and mapping.

462. How can robots ensure accurate localization and mapping in environments with highly variable humidity levels, such as agricultural greenhouses with controlled climates?

Answer: Robots use humidity sensors, environmental modeling, and adaptive navigation to compensate for variations in humidity levels, maintaining precise localization and mapping.

463. What role does human-robot collaboration play in improving robot localization and mapping accuracy in scenarios involving teamwork, such as disaster response missions?

Answer: Human-robot collaboration allows humans to provide guidance and feedback to robots, enhancing localization and mapping accuracy during teamwork in disaster response missions.

464. How do robots address localization challenges when operating in environments with rapidly changing ocean currents, such as marine exploration in open waters?

Answer: Robots use ocean current sensors, adaptive navigation strategies, and real-time updates to adapt to changing ocean currents and maintain accurate localization and mapping.

465. What techniques do robots employ for real-time mapping and localization in the field of archaeobotany, including the study of ancient plant remains and agriculture?

Answer: Robots use specialized sensors for plant analysis, archaeological data integration, and artifact recognition to optimize localization and mapping for archaeobotanical research, aiding in the study of ancient plant remains and agriculture.

466. What challenges and solutions are associated with robot localization and mapping in environments with varying levels of radio frequency interference, such as industrial facilities with wireless communication networks?

Answer: Challenges include RF signal interference. Solutions involve spectrum analysis, interference mapping, and adaptive communication protocols to mitigate RF interference effects and maintain accurate localization and mapping.

467. How do robots handle localization challenges when operating in environments with rapidly changing humidity and temperature gradients, such as rainforests with microclimates?

Answer: Robots use microclimate sensors, localized environmental modeling, and adaptive navigation to adapt to changing humidity and temperature gradients, ensuring precise localization and mapping.

468. What techniques do robots employ for real-time mapping and localization in applications involving autonomous aerial vehicles (drones), such as aerial surveys and inspections?

Answer: Robots use GPS and IMU sensors, computer vision, and aerial mapping techniques to navigate and map accurately in aerial applications, supporting surveys and inspections.

469. How can robots ensure accurate localization and mapping in environments with varying levels of terrain softness, such as sandy deserts or marshlands?

Answer: Robots use terrain analysis, adaptive traction control, and multi-sensor fusion to handle localization challenges in environments with varying terrain softness.

470. What is the significance of predictive modeling in robot localization and mapping to anticipate dynamic changes in the environment, such as water level fluctuations in rivers?

Answer: Predictive modeling allows robots to forecast environmental changes, aiding in anticipation and adaptation to dynamic environmental conditions for accurate localization and mapping.

471. How do robots address localization challenges when navigating through confined spaces with complex geometry, such as archaeological caves or ancient catacombs?

Answer: Robots use 3D mapping techniques, obstacle detection, and adaptive motion planning to navigate accurately in confined spaces with complex geometry.

472. What challenges and solutions are associated with robot localization and mapping in environments with rapid changes in air pressure and oxygen levels, such as high-altitude mountain regions?

Answer: Challenges include altitude variation. Solutions involve pressure and oxygen sensors, altitude modeling, and adaptive navigation to handle localization and mapping in high-altitude environments.

473. How can robots optimize their localization and mapping approaches for applications in precision livestock farming, including animal monitoring and feeding automation?

Answer: Robots use livestock monitoring sensors, feeding equipment integration, and precision mapping to optimize localization and mapping for precision livestock farming tasks, supporting animal monitoring and feeding automation.

474. How do robots handle localization challenges when operating in environments with rapidly changing water salinity levels, such as estuaries and coastal areas?

Answer: Robots use salinity sensors, water quality monitoring, and adaptive navigation to adapt to changing water salinity levels and maintain accurate localization and mapping.

475. What role does swarm robotics play in improving robot localization and mapping in scenarios involving multiple cooperating robots, such as disaster response missions?

Answer: Swarm robotics enables multiple robots to collaborate, share maps, and coordinate localization and mapping efforts, enhancing efficiency and accuracy in disaster response missions.

476. How can robots ensure accurate localization and mapping in environments with unpredictable seismic activity, such as earthquake-prone regions?

Answer: Robots use seismic sensors, earthquake prediction models, and adaptive navigation to adapt to seismic activity and maintain accurate localization and mapping.

477. What techniques do robots employ for real-time mapping and localization in underwater archaeology, including the exploration of submerged shipwrecks and historical artifacts?

Answer: Robots use underwater archaeological sensors, 3D scanning, and artifact recognition to navigate and map accurately in underwater archaeology, preserving historical artifacts with precision.

478. How do robots address localization challenges when navigating through environments with rapidly changing soil conditions, such as construction sites with excavation and backfilling activities?
Answer: Robots use soil condition sensors, construction site monitoring, and adaptive path planning to adapt to changing soil conditions and maintain accurate localization and mapping.
479. What is the significance of adaptive fusion of visual and inertial sensors in improving robot localization and mapping accuracy during rapid maneuvers and motion changes?
Answer: Adaptive fusion of visual and inertial sensors allows robots to maintain accurate localization and mapping during rapid motions, enhancing performance in dynamic environments.
480. How can robots optimize their localization and mapping approaches for applications in the field of marine ecology, including the study of underwater ecosystems and marine species?
Answer: Robots use underwater ecology sensors, marine species recognition tools, and underwater mapping to optimize localization and mapping for marine ecology research, aiding in the study of underwater ecosystems and species.
481. What challenges and solutions are associated with robot localization and mapping in environments with extensive human activity and interactions, such as shopping malls or amusement parks?
Answer: Challenges include crowd analysis. Solutions involve real-time crowd monitoring, pedestrian behavior modeling, and adaptive navigation to navigate and map accurately in crowded environments.
482. How do robots handle localization challenges when operating in environments with rapidly changing water currents and turbulence, such as river navigation in turbulent waters?
Answer: Robots use water current sensors, turbulence prediction algorithms, and adaptive navigation to adapt to changing water currents and turbulence, maintaining accurate localization and mapping.
483. What techniques do robots employ for real-time mapping and localization in the field of marine geophysics, including seafloor exploration and geological surveys?
Answer: Robots use seafloor geophysics sensors, geological mapping tools, and 3D imaging to navigate and map accurately in marine geophysics research, supporting seafloor exploration and geological surveys.
484. How can robots ensure accurate localization and mapping in environments with unpredictable airborne particles, such as dust storms in desert regions?
Answer: Robots use particle sensors, dust storm prediction models, and adaptive navigation to adapt to airborne particles and maintain accurate localization and mapping.
485. What is the significance of collaborative multi-robot localization and mapping in large-scale industrial environments with multiple robotic units, such as factories or warehouses?
Answer: Collaborative multi-robot localization and mapping allows multiple robots to share maps and coordinate efforts, improving efficiency and accuracy in large-scale industrial environments.
486. How do robots address localization challenges when navigating in environments with rapidly changing vegetation growth patterns, such as agricultural fields during crop cycles?
Answer: Robots use real-time crop monitoring, vegetation modeling, and adaptive path planning to handle localization challenges in environments with changing vegetation growth patterns.
487. What challenges and solutions are associated with robot localization and mapping in environments with significant variations in gravitational force, such as space exploration missions?
Answer: Challenges include gravity fluctuations. Solutions involve adaptive sensor calibration and navigation algorithms to adapt to variations in gravitational force and maintain accurate localization and mapping.
488. How can robots optimize their localization and mapping approaches for applications in the field of marine chemistry, including water quality analysis and chemical sampling?
Answer: Robots use marine chemistry sensors, water quality modeling, and chemical sampling tools to optimize localization and mapping for marine chemistry research, supporting water quality analysis and chemical sampling.
489. What role does augmented reality (AR) play in enhancing robot-human collaboration during localization and mapping tasks in mixed reality environments?
Answer: AR overlays digital information in the real world, aiding robot-human collaboration by providing real-time information and instructions during localization and mapping tasks in mixed reality environments.

490. How do robots handle localization challenges when operating in environments with rapidly changing tidal patterns, such as coastal exploration missions?
Answer: Robots use tidal prediction models, adaptive navigation, and real-time updates to adapt to changing tidal patterns and maintain accurate localization and mapping.
491. What techniques do robots employ for real-time mapping and localization in the field of marine meteorology, including weather monitoring and oceanographic data collection?
Answer: Robots use marine meteorology sensors, oceanographic instruments, and weather modeling to navigate and map accurately in marine meteorology research, supporting weather monitoring and data collection.
492. How can robots ensure accurate localization and mapping in environments with challenging acoustic conditions, such as underwater acoustic interference in sonar-based mapping?
Answer: Robots use sonar interference detection, adaptive signal processing, and underwater acoustic modeling to mitigate acoustic interference effects and maintain accurate localization and mapping.
493. What is the significance of adaptive terrain analysis in improving robot localization and mapping accuracy in environments with diverse and rugged terrain?
Answer: Adaptive terrain analysis allows robots to assess and adapt to diverse and rugged terrains, enhancing localization and mapping accuracy in challenging environments.
494. How can robots optimize their localization and mapping approaches for applications in the field of marine biology, including underwater species identification and habitat monitoring?
Answer: Robots use underwater biology sensors, species recognition algorithms, and habitat mapping to optimize localization and mapping for marine biology research, supporting species identification and habitat monitoring.
495. How do robots address localization challenges when navigating through environments with rapidly changing lighting conditions, such as forested areas with shifting sunlight patterns?
Answer: Robots use real-time lighting analysis, adaptive camera settings, and lighting correction techniques to adapt to changing lighting conditions and maintain accurate localization and mapping.
496. What challenges and solutions are associated with robot localization and mapping in environments with extensive electromagnetic interference from power lines and electrical equipment, such as substations?
Answer: Challenges include electromagnetic interference. Solutions involve electromagnetic shielding, interference mapping, and adaptive sensor calibration to mitigate interference effects and maintain accurate localization and mapping.
497. How can robots ensure accurate localization and mapping in environments with rapidly changing water temperature profiles, such as river navigation in regions with temperature gradients?
Answer: Robots use temperature sensors, water temperature modeling, and adaptive navigation to adapt to changing water temperature profiles and maintain accurate localization and mapping.
498. What role does machine learning play in improving robot perception and localization in environments with evolving object dynamics, such as industrial automation and warehouse logistics?
Answer: Machine learning allows robots to adapt to evolving object dynamics by learning from past experiences, enhancing perception and localization in dynamic environments.
499. How do robots address localization challenges when operating in environments with rapidly changing air quality due to industrial emissions, such as factory floors with pollutants?
Answer: Robots use air quality sensors, pollutant modeling, and adaptive navigation to adapt to changing air quality conditions and maintain accurate localization and mapping.
500. What is the significance of cloud-based collaborative mapping platforms in improving the scalability and accuracy of robot localization and mapping systems in large-scale collaborative efforts?
Answer: Cloud-based collaborative mapping platforms enable multiple robots to share and collaborate on maps in real-time, enhancing scalability and accuracy in large-scale localization and mapping projects involving multiple robotic units.
501. What challenges and solutions are associated with robot localization and mapping in environments with dynamic and unpredictable wind patterns, such as open-pit mining operations?

Answer: Challenges include wind interference. Solutions involve wind modeling, obstacle avoidance, and adaptive navigation to mitigate wind effects and maintain accurate localization and mapping.

502. How do robots handle localization challenges when operating in environments with rapidly changing water chemistry, such as aquatic research missions in varying water bodies?

Answer: Robots use water chemistry sensors, chemical modeling, and adaptive navigation to adapt to changing water chemistry and maintain accurate localization and mapping.

503. What techniques do robots employ for real-time mapping and localization in applications involving autonomous underwater vehicles (AUVs), such as marine exploration and seafloor mapping?

Answer: Robots use AUV-specific sensors, acoustic mapping, and underwater navigation to navigate and map accurately in underwater applications, supporting marine exploration and seafloor mapping.

504. How can robots ensure accurate localization and mapping in environments with frequent variations in air pressure due to weather changes, such as atmospheric research missions?

Answer: Robots use pressure sensors, weather forecasting data, and adaptive navigation to compensate for variations in air pressure, maintaining precise localization and mapping.

505. What is the significance of adaptive sensor calibration in improving robot localization and mapping accuracy, especially when using a variety of sensor types?

Answer: Adaptive sensor calibration allows robots to continuously adjust sensor parameters, ensuring precise and consistent sensor measurements for accurate localization and mapping in diverse sensor environments.

506. How do robots address localization challenges when navigating through environments with rapidly changing vegetation types, such as forests with seasonal foliage variations?

Answer: Robots use seasonal vegetation modeling, adaptive vision systems, and real-time path planning to handle localization challenges in environments with changing vegetation types.

507. What challenges and solutions are associated with robot localization and mapping in environments with highly reflective water surfaces, such as navigation on lakes or ponds?

Answer: Challenges include water reflection. Solutions involve polarized camera filters, water surface modeling, and adaptive algorithms to mitigate reflection effects and maintain accurate localization and mapping.

508. How can robots optimize their localization and mapping approaches for applications in the field of marine physics, including ocean current monitoring and turbulence analysis?

Answer: Robots use marine physics sensors, oceanographic instruments, and fluid dynamics modeling to optimize localization and mapping for marine physics research, supporting ocean current monitoring and turbulence analysis.

509. What role does human-robot collaboration play in improving robot localization and mapping accuracy in scenarios involving cooperative exploration, such as planetary missions?

Answer: Human-robot collaboration allows humans to provide guidance and insights, enhancing localization and mapping accuracy during cooperative exploration missions in planetary environments.

510. How do robots handle localization challenges when operating in environments with rapidly changing water acidity levels, such as aquatic research in acidic lakes or volcanic regions?

Answer: Robots use acidity sensors, water quality monitoring, and adaptive navigation to adapt to changing water acidity levels and maintain accurate localization and mapping.

511. What techniques do robots employ for real-time mapping and localization in applications involving swarm robotics, such as search and rescue missions with multiple coordinated robots?

Answer: Robots use swarm coordination algorithms, inter-robot communication, and collaborative mapping techniques to navigate and map accurately in swarm robotics applications, supporting search and rescue missions.

512. How can robots ensure accurate localization and mapping in environments with challenging acoustic conditions, such as underwater environments with complex sound propagation?

Answer: Robots use acoustic modeling, underwater acoustics sensors, and adaptive signal processing to mitigate acoustic challenges and maintain accurate localization and mapping.

513. What is the significance of terrain-aware navigation in improving robot localization and mapping accuracy in environments with complex and uneven terrains?

Answer: Terrain-aware navigation allows robots to consider the terrain's characteristics, enhancing localization and mapping accuracy in challenging environments with uneven terrain.

514. How do robots address localization challenges when navigating through environments with rapidly changing water turbidity levels, such as underwater inspections in sediment-laden waters?

Answer: Robots use turbidity sensors, adaptive vision systems, and real-time mapping updates to adapt to changing water turbidity levels and maintain accurate localization and mapping.

515. What role does swarm intelligence play in enhancing robot localization and mapping in scenarios involving multiple cooperating robots, such as agricultural field operations?

Answer: Swarm intelligence enables multiple robots to coordinate and share information, improving localization and mapping accuracy in collaborative agricultural field operations.

516. How can robots optimize their localization and mapping approaches for applications in the field of marine archaeology, including shipwreck exploration and underwater artifact recovery?

Answer: Robots use underwater archaeology sensors, shipwreck recognition algorithms, and underwater artifact mapping to optimize localization and mapping for marine archaeology research, aiding in shipwreck exploration and artifact recovery.

517. What challenges and solutions are associated with robot localization and mapping in environments with frequent interference from natural magnetic anomalies, such as regions with geological variations?

Answer: Challenges include magnetic anomaly interference. Solutions involve magnetic anomaly detection, adaptive sensor calibration, and navigation algorithms to handle interference effects and maintain accurate localization and mapping.

518. How do robots handle localization challenges when operating in environments with rapidly changing water salinity and temperature gradients, such as estuaries with tidal inflows?

Answer: Robots use multi-sensor fusion, salinity and temperature modeling, and adaptive navigation to adapt to changing water conditions and maintain accurate localization and mapping.

519. What is the significance of reinforcement learning in improving robot perception and localization in environments with dynamic and evolving object interactions, such as warehouses with changing inventory?

Answer: Reinforcement learning allows robots to adapt to evolving object interactions, enhancing perception and localization in dynamic environments like warehouses with changing inventory.

520. How can robots ensure accurate localization and mapping in environments with unpredictable gas emissions, such as volcanic regions with varying gas concentrations?

Answer: Robots use gas sensors, emission modeling, and adaptive navigation to adapt to unpredictable gas emissions and maintain accurate localization and mapping.

521. What techniques do robots employ for real-time mapping and localization in precision construction applications, such as autonomous bulldozers and excavation equipment?

Answer: Robots use construction-specific sensors, site modeling, and real-time construction monitoring to navigate and map accurately in precision construction applications, supporting autonomous equipment operation.

522. How do robots address localization challenges when navigating through environments with rapidly changing water flow patterns, such as river navigation in changing currents and eddies?

Answer: Robots use water flow sensors, hydrodynamic modeling, and adaptive navigation to adapt to changing water flow patterns and maintain accurate localization and mapping.

523. What challenges and solutions are associated with robot localization and mapping in environments with significant variations in humidity and airborne particles, such as indoor factories with dust and moisture?

Answer: Challenges include environmental variations. Solutions involve environmental sensing, indoor mapping techniques, and adaptive algorithms to handle variations in humidity and airborne particles for accurate localization and mapping.

524. How can robots optimize their localization and mapping approaches for applications in the field of marine limnology, including the study of freshwater ecosystems and water quality analysis?

Answer: Robots use limnology sensors, freshwater ecosystem mapping, and water quality assessment tools to optimize localization and mapping for marine limnology research, supporting freshwater ecosystem study and water quality analysis.

525. What role does augmented reality (AR) play in enhancing robot-human collaboration during localization and mapping tasks in mixed reality environments, such as collaborative fieldwork?

Answer: AR enhances robot-human collaboration by overlaying digital information in the real world, providing real-time guidance and information during localization and mapping tasks in mixed reality environments, promoting effective collaboration in fieldwork scenarios.

526. How do robots handle localization challenges when operating in environments with rapidly changing geological conditions, such as regions with active geological processes?

Answer: Robots use geological monitoring sensors, real-time geological analysis, and adaptive navigation to adapt to changing geological conditions and maintain accurate localization and mapping.

527. What is the significance of adaptive control systems in improving robot perception and localization during dynamic and high-speed maneuvers, such as autonomous racing competitions?

Answer: Adaptive control systems enable robots to adjust their behaviors during high-speed maneuvers, enhancing perception and localization in dynamic and fast-paced scenarios like autonomous racing competitions.

528. How can robots ensure accurate localization and mapping in environments with rapidly changing atmospheric conditions, such as weather balloons or atmospheric research missions?

Answer: Robots use atmospheric sensors, weather forecasting data, and adaptive navigation to adapt to changing atmospheric conditions and maintain accurate localization and mapping.

529. What techniques do robots employ for real-time mapping and localization in marine biogeochemistry research, including the study of nutrient cycling and oceanic carbon dynamics?

Answer: Robots use marine biogeochemistry sensors, nutrient cycling modeling, and oceanic carbon tracking to navigate and map accurately in marine biogeochemistry research, supporting the study of nutrient cycling and carbon dynamics.

530. How do robots address localization challenges when navigating through environments with rapidly changing air flow patterns, such as aerial surveillance in turbulent winds?

Answer: Robots use wind monitoring sensors, turbulence prediction algorithms, and adaptive aerial navigation to adapt to changing air flow patterns and maintain accurate localization and mapping.

531. What challenges and solutions are associated with robot localization and mapping in environments with extensive radio frequency interference from electronic devices, such as smart cities with IoT networks?

Answer: Challenges include RF interference. Solutions involve interference mapping, frequency spectrum analysis, and adaptive communication protocols to mitigate RF interference effects and maintain accurate localization and mapping.

532. How do robots handle localization challenges when operating in environments with rapidly changing terrain due to geological activities, such as volcanic eruptions or landslides?

Answer: Robots use geological monitoring, real-time hazard assessment, and adaptive navigation to adapt to changing terrain caused by geological activities, ensuring precise localization and mapping.

533. What techniques do robots employ for real-time mapping and localization in applications involving autonomous agricultural machinery, such as robotic tractors and harvesters?

Answer: Robots use agricultural sensors, field mapping algorithms, and precision farming tools to navigate and map accurately in agricultural applications, supporting autonomous machinery operation.

534. How can robots ensure accurate localization and mapping in environments with unpredictable biological interactions, such as wildlife monitoring in natural reserves?

Answer: Robots use wildlife monitoring sensors, behavior modeling, and adaptive navigation to adapt to unpredictable biological interactions and maintain accurate localization and mapping.

535. What is the significance of dynamic reconfiguration in improving robot localization and mapping accuracy when the environment undergoes rapid changes, such as structural collapses in disaster scenarios?

Answer: Dynamic reconfiguration allows robots to adapt their strategies and sensor configurations in response to rapid environmental changes, enhancing localization and mapping accuracy in dynamic and evolving situations.

536. How do robots address localization challenges when navigating through environments with rapidly changing water flow velocities, such as river navigation during flooding events?

Answer: Robots use water velocity sensors, flood modeling, and adaptive navigation to adapt to changing water flow velocities and maintain accurate localization and mapping.

537. What role does swarm robotics play in improving robot localization and mapping in scenarios involving multiple cooperating robots, such as agricultural crop monitoring and harvesting?

Answer: Swarm robotics enables multiple robots to collaborate, share data, and coordinate localization and mapping efforts, enhancing efficiency and accuracy in agricultural tasks like crop monitoring and harvesting.

538. How can robots optimize their localization and mapping approaches for applications in the field of marine sedimentology, including the study of seafloor sediments and geological history?

Answer: Robots use sedimentology sensors, seafloor sediment mapping techniques, and geological analysis tools to optimize localization and mapping for marine sedimentology research, supporting the study of seafloor sediments and geological history.

539. What challenges and solutions are associated with robot localization and mapping in environments with rapid changes in atmospheric pressure, such as high-altitude scientific balloon missions?

Answer: Challenges include rapid pressure changes. Solutions involve pressure compensation mechanisms, real-time pressure monitoring, and adaptive navigation to handle pressure variations and maintain accurate localization and mapping.

540. How do robots handle localization challenges when operating in environments with rapidly changing ice conditions, such as Polar Regions with shifting ice floes?

Answer: Robots use ice condition sensors, ice flow modeling, and adaptive navigation to adapt to changing ice conditions and maintain accurate localization and mapping.

541. What techniques do robots employ for real-time mapping and localization in marine geotechnical engineering research, including seabed soil analysis and foundation design?

Answer: Robots use geotechnical sensors, seabed soil sampling tools, and foundation modeling to navigate and map accurately in marine geotechnical engineering research, supporting seabed soil analysis and foundation design.

542. How can robots ensure accurate localization and mapping in environments with challenging magnetic anomalies, such as areas with buried metallic objects?

Answer: Robots use magnetic anomaly detection, interference compensation techniques, and adaptive sensor calibration to mitigate magnetic interference effects and maintain accurate localization and mapping.

543. What is the significance of adaptive perception fusion in improving robot localization and mapping accuracy when using multiple sensory modalities, such as vision, LiDAR, and radar?

Answer: Adaptive perception fusion allows robots to integrate and reconcile data from multiple sensory modalities, enhancing localization and mapping accuracy by leveraging the strengths of each sensor type.

544. How do robots address localization challenges when navigating through environments with rapidly changing light spectra, such as underwater exploration in colorful coral reefs?

Answer: Robots use color adaptation algorithms, underwater spectral analysis, and real-time image processing to adapt to changing light spectra and maintain accurate localization and mapping.

545. What challenges and solutions are associated with robot localization and mapping in environments with extensive wireless communication networks, such as smart cities with 5G connectivity?

Answer: Challenges include network interference. Solutions involve network mapping, interference analysis, and adaptive communication protocols to mitigate network interference effects and maintain accurate localization and mapping.

546. How can robots optimize their localization and mapping approaches for applications in the field of marine paleontology, including the study of ancient marine life and fossils?

Answer: Robots use paleontology sensors, fossil recognition algorithms, and underwater fossil mapping to optimize localization and mapping for marine paleontology research, aiding in the study of ancient marine life and fossils.

547. What role does predictive analytics play in enhancing robot localization and mapping in environments with dynamic and evolving conditions, such as precision agriculture during changing weather patterns?
Answer: Predictive analytics allows robots to anticipate and adapt to evolving environmental conditions, enhancing localization and mapping in dynamic scenarios like precision agriculture.
548. How do robots handle localization challenges when operating in environments with rapidly changing water acidity and alkalinity levels, such as aquatic research in volcanic lakes?
Answer: Robots use pH sensors, chemical modeling, and adaptive navigation to adapt to changing water acidity and alkalinity levels and maintain accurate localization and mapping.
549. What techniques do robots employ for real-time mapping and localization in applications involving autonomous submersibles, such as deep-sea exploration and hydrothermal vent analysis?
Answer: Robots use submersible-specific sensors, deep-sea mapping techniques, and hydrothermal vent research tools to navigate and map accurately in underwater applications, supporting deep-sea exploration and vent analysis.
550. How can robots ensure accurate localization and mapping in environments with unpredictable electromagnetic interference, such as radio telescope sites with sensitive equipment?
Answer: Robots use interference monitoring, shielding techniques, and adaptive sensor calibration to mitigate electromagnetic interference effects and maintain accurate localization and mapping.

ONLINE RESOURCES

Practice Paper Link: https://forms.gle/q1qphZ2ZBJ9vZrAE6

QR Code:

Video Content Link:
https://www.youtube.com/playlist?list=PLzlh_31VWEFp20UnKz260nT6nLyqUyBTn

QR Code:

Chapter 9: Human-Robot Interaction

1. How does a robot's design influence its approachability to humans?

Answer: A robot's design, including its form, size, and appearance, significantly impacts its approachability, as a friendly and non-threatening design tends to make humans more comfortable around it.

2. What role does natural language processing play in enhancing human-robot communication?

Answer: Natural language processing enables robots to understand and respond to human speech, facilitating seamless verbal interactions between humans and robots.

3. How can robots exhibit empathy in their interactions with humans?

Answer: Robots can demonstrate empathy by recognizing and responding to human emotions, showing understanding, and offering appropriate support or feedback.

4. What are the benefits of using haptic feedback in human-robot interactions?

Answer: Haptic feedback allows humans to perceive touch and pressure from robots, enhancing the sense of presence and enabling more intuitive interactions in tasks like remote surgery or virtual collaboration.

5. How do robots maintain user privacy and data security during interactions that involve personal information?

Answer: Robots ensure user privacy and data security through encryption, secure communication protocols, and strict adherence to privacy regulations, safeguarding personal information shared during interactions.

6. What is the significance of anticipatory behavior in human-robot interactions?

Answer: Anticipatory behavior enables robots to predict human needs and preferences, proactively assisting or responding to users' requests, enhancing the overall user experience.

7. How can robots adapt their communication styles to suit the preferences and cultural backgrounds of different users?

Answer: Robots can adapt their communication styles by analyzing user behavior and feedback, adjusting language, gestures, and expressions to align with individual preferences and cultural norms.

8. What ethical considerations are essential when designing robots for human interaction?

Answer: Ethical considerations include transparency in robot capabilities, informed consent, respect for autonomy, and addressing bias in AI systems to ensure ethical and responsible human-robot interactions.

9. How do robots use computer vision to interpret and respond to human facial expressions?

Answer: Robots utilize computer vision algorithms to recognize facial expressions, allowing them to assess human emotions and respond accordingly, promoting empathetic interactions.

10. What strategies do robots employ to ensure safety during physical interactions with humans in collaborative tasks?

Answer: Robots implement safety features such as force sensors, collision detection, and compliance control to minimize the risk of harm during physical interactions, prioritizing human safety.

11. How can robots assist individuals with disabilities in improving their quality of life through tailored interactions?

Answer: Robots can provide personalized assistance, including mobility support, communication aids, and adaptive interfaces, to empower individuals with disabilities and enhance their daily lives.

12. What is the role of emotional intelligence in human-robot interactions, and how can robots develop it?

Answer: Emotional intelligence enables robots to recognize and respond to human emotions sensitively. Robots can develop it through machine learning models trained on emotional data and human feedback.

13. How do robots manage user expectations and avoid overpromising capabilities during interactions?

Answer: Robots employ honest communication and set clear expectations about their capabilities to avoid overpromising, ensuring users have realistic perceptions of what the robot can achieve.

14. What are the challenges and benefits of multi-modal communication in human-robot interactions?

Answer: Multi-modal communication combines various sensory modalities (e.g., speech, gestures, and facial expressions) for richer interactions. Challenges include synchronization, but benefits include improved understanding and engagement.

15. How can robots use gesture recognition to facilitate intuitive and non-verbal communication with humans?

Answer: Robots employ gesture recognition algorithms to interpret and generate gestures, allowing for natural non-verbal communication, enhancing understanding, and engagement.

16. What safeguards are in place to prevent robots from misinterpreting or misusing personal information shared during conversations?

Answer: Safeguards include data encryption, access controls, and strict data handling protocols to ensure robots protect and respect user privacy during conversations.

17. How do robots maintain social boundaries and appropriate behavior during interactions, particularly in sensitive or emotional situations?

Answer: Robots follow predefined social rules and behavior guidelines to maintain appropriate conduct during interactions, respecting cultural and emotional sensitivities.

18. What role does explainability play in building trust between humans and robots during decision-making processes?

Answer: Explainability allows robots to provide transparent explanations of their decisions, enabling humans to understand and trust the reasoning behind robot actions.

19. How can robots facilitate remote collaboration between individuals in different locations, and what technologies support this?

Answer: Robots support remote collaboration through telepresence capabilities, enabling real-time video, audio, and interaction between remote users, fostering collaboration across distances.

20. What strategies do robots use to adapt to the emotional state of their users, such as calming anxious individuals or cheering up someone feeling down?

Answer: Robots adapt to emotional states through sentiment analysis and responsive behavior, offering comforting words or activities to calm anxious or uplift despondent users.

21. How do robots ensure accessibility for individuals with diverse needs, including those with sensory impairments?

Answer: Robots provide accessibility features such as screen readers, sign language interpretation, and alternative input methods to accommodate individuals with sensory impairments.

22. What is the role of feedback loops in continuous improvement of human-robot interactions?

Answer: Feedback loops involve gathering user feedback, analyzing it, and making iterative improvements to robot behavior and communication, leading to better user experiences over time.

23. How can robots enhance learning experiences by personalizing content and adapting to individual learning styles?

Answer: Robots use AI-driven personalization to tailor educational content and methods to match individual learning styles, optimizing the learning process for each user.

24. What safeguards are in place to prevent robots from unintentionally promoting biased or discriminatory behaviors in their interactions?

Answer: Safeguards include bias detection algorithms and ethical guidelines to ensure robots avoid promoting biased or discriminatory behaviors, fostering fair and inclusive interactions.

25. How do robots handle situations where user instructions conflict with ethical or legal guidelines, ensuring responsible behavior?

Answer: Robots follow ethical and legal guidelines, refusing instructions that conflict with them and offering alternative solutions or advice when necessary.

26. What measures are in place to prevent robots from becoming excessively dependent on human interaction and maintaining autonomy?

Answer: Robots use autonomy preservation mechanisms to balance human interaction with independent decision-making, avoiding overreliance on human input.

27. How can robots support mental health and well-being by providing companionship and emotional support to users?

Answer: Robots offer companionship and emotional support through empathetic conversations, active listening, and the provision of coping strategies, benefiting users' mental health.

28. What strategies do robots employ to engage children and enhance their learning experiences in educational settings?

Answer: Robots engage children through gamified learning, interactive storytelling, and age-appropriate content, fostering enjoyable and educational experiences in classrooms.

29. How do robots address cultural sensitivities and differences in communication norms during international interactions?

Answer: Robots use cultural sensitivity training and adapt communication styles to align with the cultural norms and preferences of users from different regions and backgrounds.

30. How can robots assist in healthcare settings by providing reminders, monitoring vital signs, and offering emotional support to patients?

Answer: Robots play a role in healthcare by reminding patients of medication schedules, monitoring vital signs, and providing emotional support, contributing to patient well-being and recovery.

31. What measures do robots take to ensure user safety when they operate in shared spaces with humans, such as collaborative manufacturing environments?

Answer: Robots incorporate safety features like proximity sensors, collision avoidance, and speed reduction in shared spaces. They follow safety standards and protocols to minimize risks to humans during collaborative operations.

32. How can robots assist in language learning and language acquisition for individuals seeking to acquire new languages or improve their language skills?

Answer: Robots support language learning through interactive language lessons, pronunciation correction, and vocabulary-building exercises, offering personalized and effective language acquisition tools.

33. What role does emotion recognition technology play in enabling robots to respond appropriately to the emotional states of their users?

Answer: Emotion recognition technology allows robots to detect user emotions and adjust their responses accordingly, tailoring interactions to provide emotional support, comfort, or assistance.

34. How do robots address privacy concerns when they perform tasks like surveillance or home monitoring, where they may capture sensitive information?

Answer: Robots prioritize privacy by enabling data encryption, anonymizing footage, and providing user control over data access, ensuring responsible use in surveillance and monitoring applications.

35. What strategies do robots use to enhance accessibility for individuals with cognitive disabilities, such as autism or dementia?

Answer: Robots employ visual aids, simplified language, and structured routines to assist individuals with cognitive disabilities, helping them navigate daily tasks and social interactions more comfortably.

36. How can robots contribute to improving mental health by offering therapeutic interventions, stress relief techniques, or companionship?

Answer: Robots offer mental health support through guided relaxation exercises, emotional conversations, and companionship, aiding in stress reduction and emotional well-being.

37. What measures do robots implement to prevent security breaches and unauthorized access to sensitive data during human-robot interactions?

Answer: Robots employ robust authentication, encryption, and access control mechanisms to safeguard sensitive data, ensuring that unauthorized users cannot compromise security.

38. How do robots foster inclusivity in educational settings by accommodating diverse learning needs and abilities?

Answer: Robots support inclusivity in education by offering adaptable teaching methods, providing resources for special needs students, and accommodating various learning styles and abilities.

39. What ethical considerations are essential when deploying robots in caregiving roles, such as assisting the elderly or individuals with disabilities?
Answer: Ethical considerations include informed consent, respecting autonomy, preserving dignity, and ensuring that robots do not replace essential human care and compassion in caregiving contexts.
40. How can robots contribute to disaster response efforts by assisting in search and rescue missions or providing critical information during crises?
Answer: Robots aid in disaster response by navigating hazardous environments, locating survivors, and relaying critical information to first responders, enhancing the effectiveness of search and rescue operations.
41. What mechanisms do robots employ to recognize and respond to non-verbal cues, such as body language and facial expressions, in human interactions?
Answer: Robots use computer vision and machine learning to interpret non-verbal cues, enabling them to understand emotions and intentions conveyed through body language and facial expressions, promoting more natural interactions.
42. How can robots support elderly individuals in aging populations by assisting with daily tasks, medication reminders, and companionship?
Answer: Robots offer companionship and assistance to elderly individuals by helping with tasks, providing medication reminders, and engaging in conversation, thereby enhancing the quality of life for aging populations.
43. What measures do robots take to maintain user trust and transparency in decision-making processes, particularly in complex tasks like autonomous vehicles?
Answer: Robots ensure transparency by explaining their decision logic in real-time, allowing users to understand and trust their actions, which is crucial for safe and confident interactions, especially in autonomous vehicles.
44. How do robots address challenges related to cultural and linguistic diversity when interacting with users from different backgrounds?
Answer: Robots adapt to cultural diversity by offering multilingual support, recognizing cultural norms, and accommodating communication preferences, facilitating meaningful interactions across diverse user groups.
45. How can robots assist in education by providing personalized tutoring, adapting to individual learning paces, and tracking progress?
Answer: Robots enhance education by tailoring lessons to individual needs, offering adaptive tutoring, and monitoring student progress, thereby improving learning outcomes and engagement.
46. What safeguards are in place to protect children from potentially harmful or inappropriate content in robot-assisted educational environments?
Answer: Safeguards include content filtering, parental controls, and age-appropriate content selection to ensure that children are exposed only to educational and safe materials during robot-assisted learning.
47. What role do affective computing technologies play in enabling robots to recognize and respond to user emotions during interactions?
Answer: Affective computing enables robots to analyze user emotions through speech, facial expressions, and physiological cues, allowing them to provide empathetic responses and support during emotional moments.
48. How do robots manage user data and ensure compliance with data privacy regulations when collecting and processing information during interactions?
Answer: Robots adhere to data privacy regulations by securely storing and processing user data, obtaining informed consent, and providing transparency regarding data usage, ensuring legal and ethical practices.
49. What strategies do robots employ to assist individuals with neurodiversity conditions, such as autism or ADHD, in improving their social skills and daily routines?
Answer: Robots assist neurodiversity individuals through social skill coaching, routine management, and sensory support, helping them navigate social interactions and daily life more effectively.
50. How can robots contribute to environmental sustainability by assisting in tasks like waste management, energy conservation, or monitoring pollution levels?
Answer: Robots promote sustainability by automating eco-friendly practices, such as efficient waste sorting, energy-efficient processes, and real-time environmental monitoring, contributing to a cleaner and greener future.

51. What measures do robots take to ensure the physical safety of users during interactions, particularly in healthcare or physical therapy applications?
Answer: Robots prioritize user safety by monitoring physical conditions, implementing safety protocols, and employing adaptive techniques to prevent accidents or injuries during healthcare interactions.
52. How can robots assist in disaster preparedness by simulating emergency scenarios, providing evacuation guidance, or remotely assessing disaster-stricken areas?
Answer: Robots aid in disaster preparedness by conducting realistic emergency drills, offering evacuation guidance, and remotely assessing disaster-affected regions, enhancing preparedness and response efforts.
53. What ethical guidelines and algorithms do robots follow to ensure fair and unbiased decision-making, especially in contexts like recruitment or loan approval?
Answer: Robots adhere to ethical guidelines and employ bias detection algorithms to prevent discrimination, ensuring that decisions related to recruitment or loan approval are fair and unbiased.
54. How do robots use gesture recognition to enhance accessibility for individuals with motor impairments, enabling them to control devices and communicate effectively?
Answer: Robots employ gesture recognition to interpret subtle movements, allowing individuals with motor impairments to control devices, communicate, and interact with the world more comfortably.
55. What measures are in place to protect user data from cybersecurity threats and hacking attempts during human-robot interactions?
Answer: Robots employ robust cybersecurity measures, including encryption, firewalls, and intrusion detection systems, to safeguard user data and prevent unauthorized access or breaches.
56. How can robots assist individuals with mental health challenges by offering therapeutic interventions, such as cognitive-behavioral therapy or mindfulness exercises?
Answer: Robots support mental health by providing therapeutic interventions like cognitive-behavioral therapy exercises, mindfulness sessions, and mood tracking, offering valuable tools for managing mental health.
57. How do robots ensure equitable access to education and information for individuals in underserved or remote communities, where resources may be limited?
Answer: Robots promote equitable access by delivering educational content, medical information, and communication tools to underserved or remote communities, bridging gaps in access to essential services.
58. What role do robots play in promoting STEM (Science, Technology, Engineering, and Mathematics) education by offering hands-on learning experiences and programming challenges?
Answer: Robots engage students in STEM education through hands-on activities, coding challenges, and interactive lessons, fostering interest and proficiency in STEM fields.
59. How can robots assist in improving workplace safety by monitoring hazardous environments, detecting potential risks, and issuing alerts to prevent accidents?
Answer: Robots enhance workplace safety by continuously monitoring environments for hazards, detecting risks, and alerting workers to potential dangers, reducing the likelihood of accidents.

60. How do robots contribute to space exploration by conducting autonomous missions, collecting scientific data, and assisting astronauts in extraterrestrial environments?
Answer: Robots play a crucial role in space exploration by conducting autonomous missions, performing experiments, and supporting astronauts with tasks like repairs and sample collection in challenging extraterrestrial environments.
61. What measures are in place to ensure the ethical use of robots in military applications, including adherence to international laws and ethical guidelines?
Answer: Robots in military applications adhere to strict ethical guidelines and international laws, emphasizing proportionality, discrimination, and minimizing harm to civilians during armed conflicts.
62. How can robots assist in addressing food security challenges by optimizing agricultural processes, monitoring crop health, and reducing post-harvest losses?

Answer: Robots contribute to food security by optimizing farming practices, identifying crop diseases early, and automating post-harvest handling, leading to increased food production and reduced waste.

63. What ethical considerations guide robots in healthcare settings, particularly in situations where they assist in sensitive medical procedures or decision-making?

Answer: Ethical considerations include patient consent, preserving dignity, and adherence to medical ethics principles when robots assist in sensitive medical procedures or decision-making processes.

64. How do robots support individuals with visual impairments by providing navigation assistance, reading printed text, and recognizing objects in their surroundings?

Answer: Robots assist visually impaired individuals through navigation aids, text-to-speech technology, and object recognition, facilitating greater independence and accessibility.

65. What strategies do robots use to facilitate cross-cultural understanding and communication, especially in international business or diplomacy contexts?

Answer: Robots promote cross-cultural understanding by providing real-time translation, cultural sensitivity training, and adaptability to diverse communication styles, fostering effective international interactions.

66. How can robots assist individuals with speech and communication disorders, such as aphasia, by providing speech therapy exercises and communication support?

Answer: Robots aid individuals with speech disorders through speech therapy exercises, augmentative and alternative communication (AAC) support, and articulation practice, facilitating improved communication skills and confidence.

67. What ethical frameworks guide robots in ensuring transparency, accountability, and fairness in decision-making processes that impact human lives, such as autonomous vehicles?

Answer: Ethical frameworks for autonomous vehicles prioritize transparency, safety, and ethical decision-making, ensuring that decisions prioritize human safety and adhere to societal values.

68. How do robots contribute to wildlife conservation efforts by monitoring endangered species, tracking animal behaviors, and assisting in anti-poaching activities?

Answer: Robots assist in wildlife conservation through wildlife monitoring, behavior analysis, and anti-poaching operations, aiding in the protection and preservation of endangered species.

69. What measures are in place to prevent robots from becoming overly persuasive or manipulative in their interactions, especially in persuasive marketing or sales roles?

Answer: Measures include guidelines against manipulation, ethical marketing practices, and transparency in communication, ensuring that robots maintain ethical boundaries in persuasive roles.

70. How can robots aid in reducing air pollution and mitigating environmental damage by monitoring air quality, identifying pollution sources, and suggesting eco-friendly practices?

Answer: Robots contribute to environmental protection by monitoring air quality, pinpointing pollution sources, and recommending eco-friendly solutions, aiding in the reduction of air pollution and environmental damage.

71. What ethical guidelines and decision-making algorithms guide robots in situations where they assist in medical diagnoses, treatment recommendations, or surgery?

Answer: Ethical guidelines and medical decision-making algorithms ensure that robots prioritize patient well-being, adhere to medical ethics, and provide accurate diagnoses and treatment recommendations in healthcare settings.

72. How do robots support the development of smart cities by monitoring infrastructure, managing traffic flow, and enhancing energy efficiency?

Answer: Robots aid in smart city development by monitoring infrastructure health, optimizing traffic management, and implementing energy-efficient solutions, contributing to urban sustainability.

73. How can robots assist in disaster recovery efforts by performing tasks like debris removal, building assessment, and emergency communication support?

Answer: Robots support disaster recovery through debris clearing, structural assessment, and communication assistance, expediting the restoration of affected areas and aiding in emergency response.

74. What measures do robots take to ensure user data privacy and security during interactions in the era of IoT (Internet of Things) and connected devices?

Answer: Robots employ robust security protocols, data encryption, and user consent mechanisms to protect user data and privacy when interacting in IoT environments.

75. How do robots enhance the accessibility of cultural heritage sites, museums, and historical locations by offering guided tours and information to visitors with diverse needs?

Answer: Robots improve cultural heritage accessibility by providing guided tours, multilingual information, and tailored experiences for visitors with varying abilities, ensuring inclusive cultural exploration.

76. What role do robots play in supporting astronauts during long-duration space missions by assisting with daily tasks, conducting experiments, and monitoring spacecraft systems?

Answer: Robots assist astronauts by managing spacecraft systems, performing experiments, and aiding with daily tasks during long-duration space missions, ensuring the success and safety of space exploration.

77. How can robots contribute to wildlife rehabilitation and rescue efforts by providing medical care, monitoring animal health, and assisting in habitat restoration?

Answer: Robots aid in wildlife rehabilitation by offering medical care, health monitoring, and habitat restoration support, facilitating the recovery and reintegration of injured or endangered animals.

78. What ethical principles guide robots in ensuring fairness and impartiality in legal processes, including legal research, case analysis, and dispute resolution?

Answer: Ethical principles ensure fairness and impartiality in legal processes by promoting unbiased research, objective case analysis, and adherence to legal ethics, promoting just outcomes.

79. How do robots support individuals with mobility challenges by providing mobility assistance, transportation services, and accessible navigation in urban environments?

Answer: Robots assist individuals with mobility challenges through mobility aids, accessible transportation, and navigation support, enhancing their mobility and independence in urban settings.

80. What measures are in place to prevent robots from infringing upon intellectual property rights or engaging in copyright violations when creating or generating content?

Answer: Measures include content filtering, copyright compliance algorithms, and adherence to copyright laws to prevent robots from infringing upon intellectual property rights when creating or generating content.

81. How can robots assist in addressing the digital divide by providing technology access, digital literacy training, and online resources to underserved communities?

Answer: Robots bridge the digital divide by offering technology access, digital literacy education, and online resources to underserved communities, empowering them with digital skills and resources.

82. What ethical guidelines and algorithms guide robots in maintaining user confidentiality and professional boundaries in therapeutic roles, such as counseling or psychotherapy?

Answer: Ethical guidelines ensure user confidentiality and professional boundaries in therapeutic roles, with algorithms promoting ethical behavior and adherence to therapeutic principles.

83. How do robots support agricultural sustainability by optimizing irrigation, crop monitoring, and pest control, leading to improved yields and reduced resource consumption?

Answer: Robots contribute to agricultural sustainability by automating irrigation, monitoring crops, and implementing precise pest control, resulting in higher yields and reduced environmental impact.

84. What strategies do robots employ to assist individuals with neurodegenerative conditions, such as Parkinson's disease, in managing symptoms and enhancing their quality of life?

Answer: Robots assist individuals with neurodegenerative conditions by providing symptom management tools, offering physical therapy exercises, and facilitating daily activities, improving their quality of life.

85. How can robots play a role in mitigating the effects of climate change by monitoring carbon emissions, conducting environmental research, and supporting sustainable practices?

Answer: Robots aid in climate change mitigation by monitoring carbon emissions, conducting environmental research, and promoting sustainable practices, contributing to efforts to combat climate change.

86. What measures do robots take to ensure user consent and data control when providing personalized recommendations, such as product suggestions or content recommendations?

Answer: Robots obtain user consent, provide data control options, and employ privacy settings to ensure users have control over the personalization of recommendations, respecting their preferences.

87. How do robots support individuals with hearing impairments by providing sign language interpretation, speech-to-text conversion, and communication assistance?

Answer: Robots assist individuals with hearing impairments by offering sign language interpretation, speech-to-text conversion, and communication support, facilitating effective interactions and inclusivity.

88. What ethical principles and guidelines guide robots in journalism and news reporting roles to ensure accuracy, fairness, and responsible information dissemination?

Answer: Ethical principles in journalism guide robots to prioritize accuracy, fairness, and responsible information dissemination, adhering to journalistic integrity and upholding public trust.

89. How can robots contribute to sustainable transportation solutions by optimizing traffic flow, promoting electric vehicles, and reducing congestion in urban areas?

Answer: Robots promote sustainable transportation through traffic management, electric vehicle support, and congestion reduction, enhancing urban mobility and environmental sustainability.

90. What role do robots play in supporting scientific research by conducting experiments, collecting data in extreme environments, and assisting with data analysis?

Answer: Robots assist scientific research by conducting experiments, gathering data in challenging environments, and aiding in data analysis, accelerating scientific discovery and exploration.

91. What measures are in place to prevent robots from engaging in harmful behaviors or promoting violence when used in entertainment or gaming applications?

Answer: Measures include content moderation, age-appropriate restrictions, and adherence to ethical guidelines to prevent robots from engaging in harmful or violent behaviors in entertainment and gaming.

92. How can robots assist in promoting gender equality and diversity in various fields, such as STEM, by offering mentorship, career guidance, and support to underrepresented individuals?

Answer: Robots promote gender equality and diversity by providing mentorship, career guidance, and support to underrepresented individuals, fostering inclusivity and diversity in various fields, including STEM.

93. How do robots support individuals with mental health conditions by offering therapeutic interventions, tracking mood changes, and connecting them with mental health professionals?

Answer: Robots assist individuals with mental health conditions by providing therapeutic interventions, mood tracking, and facilitating connections with mental health professionals, improving mental well-being.

94. What ethical principles guide robots in their roles as educators, ensuring that they prioritize the educational needs and well-being of students?

Answer: Ethical principles in education guide robots to prioritize student needs, provide accurate information, and foster a positive and supportive learning environment.

95. How can robots assist in addressing the challenges of aging populations by providing healthcare monitoring, medication reminders, and companionship to the elderly?

Answer: Robots support aging populations by monitoring health, offering medication reminders, and providing companionship, improving the overall well-being and quality of life for the elderly.

96. What measures are in place to prevent robots from engaging in discriminatory practices or reinforcing bias when involved in decision-making processes, such as hiring or loan approvals?

Answer: Measures include bias detection algorithms, ethical guidelines, and fairness checks to prevent robots from engaging in discriminatory practices and promote fair decision-making.

97. How do robots enhance the accessibility of public spaces and facilities for individuals with disabilities, such as offering autonomous wheelchair navigation or accessible building information?

Answer: Robots improve accessibility by offering autonomous wheelchair navigation, accessible building information, and assistance to individuals with disabilities, ensuring equal access to public spaces.

98. What ethical considerations guide robots in their roles as companions and caregivers to ensure that they prioritize the emotional and psychological well-being of users?

Answer: Ethical considerations emphasize the emotional and psychological well-being of users when robots act as companions and caregivers, focusing on empathy, dignity, and user comfort.

99. How can robots assist in preserving cultural heritage by digitizing historical artifacts, conducting archaeological research, and supporting conservation efforts?

Answer: Robots aid in preserving cultural heritage by digitizing artifacts, conducting archaeological research, and assisting in conservation activities, safeguarding cultural treasures for future generations.

100. What role do robots play in promoting sustainable energy practices by monitoring energy consumption, optimizing energy use, and supporting renewable energy solutions?

Answer: Robots promote sustainable energy practices by monitoring energy consumption, optimizing usage, and facilitating the adoption of renewable energy sources, contributing to a more sustainable future.

101. How can robots assist in disaster prevention by monitoring environmental conditions, issuing early warnings, and facilitating evacuation procedures?

Answer: Robots aid in disaster prevention by constantly monitoring environmental conditions, providing early warnings, and assisting in evacuations, reducing the risk and impact of natural disasters on communities.

102. What ethical principles guide robots in assisting law enforcement agencies, such as ensuring transparency, accountability, and adherence to human rights during surveillance or investigations?

Answer: Ethical principles in law enforcement ensure that robots maintain transparency, accountability, and respect for human rights during surveillance and investigative activities, upholding legal and ethical standards.

103. How do robots enhance accessibility in public transportation by offering services such as autonomous taxis, wheelchair-accessible vehicles, and route optimization for individuals with disabilities?

Answer: Robots improve accessibility in public transportation by providing autonomous taxis, accessible vehicles, and optimized routes for individuals with disabilities, making transportation more inclusive.

104. What measures are in place to prevent robots from engaging in harmful behaviors or promoting addictive content when used in entertainment or social media applications?

Answer: Measures include content moderation, age-appropriate restrictions, and adherence to ethical guidelines to prevent robots from engaging in harmful behaviors or promoting addictive content in entertainment and social media.

105. How can robots contribute to sustainable agriculture by supporting precision farming, soil monitoring, and efficient water usage in crop cultivation?

Answer: Robots support sustainable agriculture through precision farming, soil health monitoring, and water-efficient practices, promoting sustainable crop cultivation and resource conservation.

106. What ethical considerations guide robots in the field of journalism, ensuring unbiased reporting, fact-checking, and responsible information dissemination?

Answer: Ethical considerations in journalism guide robots to prioritize unbiased reporting, rigorous fact-checking, and responsible information dissemination, maintaining journalistic integrity and accuracy.

107. How do robots assist individuals with cognitive impairments, such as dementia, in daily tasks, memory support, and maintaining routines to improve their quality of life?

Answer: Robots assist individuals with cognitive impairments by providing daily task support, memory aids, and routine management, enhancing their quality of life and independence.

108. What role do robots play in promoting energy conservation by managing home appliances, optimizing energy usage, and providing real-time energy consumption feedback to users?

Answer: Robots promote energy conservation by managing home appliances, optimizing energy usage, and offering real-time energy consumption feedback to users, encouraging efficient practices.

109. What measures are in place to ensure ethical behavior and user safety when robots are used in autonomous military applications, including unmanned aerial vehicles (UAVs)?

Answer: Measures include adherence to international laws, strict rules of engagement, and ethical guidelines to ensure ethical behavior and user safety in autonomous military applications.

110. How can robots support space colonization efforts by assisting in habitat construction, resource extraction, and sustainability practices on distant planets or celestial bodies?

Answer: Robots aid in space colonization by participating in habitat construction, resource harvesting, and sustainable practices on distant planets or celestial bodies, facilitating long-term human presence.

111. What ethical frameworks guide robots in the field of legal advisory, ensuring legal compliance, client confidentiality, and unbiased legal advice?

Answer: Ethical frameworks in legal advisory guide robots to prioritize legal compliance, client confidentiality, and impartial legal advice, upholding professional ethics in the legal field.

112. How do robots enhance accessibility for individuals with visual impairments by providing audio descriptions, indoor navigation, and object recognition in public spaces?

Answer: Robots enhance accessibility for individuals with visual impairments by offering audio descriptions, indoor navigation assistance, and object recognition, improving independence and mobility.

113. What role do robots play in supporting scientific research in extreme environments, such as deep-sea exploration or polar expeditions, by conducting experiments and data collection?

Answer: Robots support scientific research in extreme environments by conducting experiments, collecting data, and performing tasks that are challenging or dangerous for human researchers.

114. What measures are in place to ensure data accuracy and privacy when robots are used in healthcare applications, such as medical diagnoses or telemedicine?

Answer: Measures include data encryption, secure transmission, and adherence to healthcare privacy regulations to ensure data accuracy and privacy in healthcare applications.

115. How can robots contribute to wildlife habitat restoration by planting trees, removing invasive species, and supporting ecosystem regeneration in degraded areas?

Answer: Robots contribute to wildlife habitat restoration by planting trees, eliminating invasive species, and aiding ecosystem regeneration, promoting biodiversity and ecological health.

116. What ethical principles guide robots in their roles as teachers and educators, ensuring that they provide fair assessments, encourage critical thinking, and support diverse learning styles?

Answer: Ethical principles in education guide robots to provide fair assessments, foster critical thinking, and support diverse learning styles, promoting equitable education.

117. How do robots assist individuals with physical disabilities by providing mobility support, assistive devices, and adaptive technology to improve their daily lives?

Answer: Robots assist individuals with physical disabilities by offering mobility support, assistive devices, and adaptive technology, enhancing their independence and quality of life.

118. What measures are in place to prevent robots from infringing upon intellectual property rights or engaging in plagiarism when generating content or conducting research?

Answer: Measures include plagiarism detection algorithms, copyright compliance checks, and adherence to intellectual property laws to prevent robots from infringing upon intellectual property rights.

119. How can robots contribute to the conservation of marine ecosystems by monitoring ocean health, conducting underwater research, and assisting in coral reef restoration efforts?

Answer: Robots contribute to marine ecosystem conservation by monitoring ocean health, conducting underwater research, and participating in coral reef restoration projects, protecting marine biodiversity.

120. What ethical guidelines and algorithms guide robots in assisting individuals with addiction by providing support, recovery resources, and harm reduction strategies?

Answer: Ethical guidelines ensure that robots prioritize addiction support, recovery resources, and harm reduction strategies while adhering to ethical principles and user well-being.

121. How do robots support sustainable transportation solutions in urban areas by optimizing public transit systems, reducing traffic congestion, and promoting eco-friendly commuting options?

Answer: Robots support sustainable transportation by optimizing public transit, reducing traffic congestion, and encouraging eco-friendly commuting options, contributing to urban sustainability.

122. What role do robots play in promoting diversity and inclusion in the workplace by offering diversity training, bias mitigation, and inclusive leadership development?

Answer: Robots promote diversity and inclusion by providing diversity training, bias mitigation, and leadership development programs, fostering inclusive workplaces and equitable opportunities.

123. How can robots assist in improving mental health outcomes by offering therapy sessions, stress management tools, and mental health awareness campaigns?

Answer: Robots contribute to mental health improvement by providing therapy sessions, stress management tools, and mental health awareness campaigns, supporting overall well-being.

124. What measures are in place to ensure responsible use of robots in law enforcement and security roles, including adherence to ethical standards and human rights principles?

Answer: Measures include adherence to ethical standards, human rights principles, and oversight mechanisms to ensure the responsible use of robots in law enforcement and security.

125. How do robots enhance accessibility in educational settings by offering adaptive learning materials, inclusive classroom support, and personalized education plans?

Answer: Robots enhance educational accessibility by providing adaptive materials, inclusive classroom support, and personalized education plans, catering to diverse learning needs.

ONLINE RESOURCES

Practice Paper Link: https://forms.gle/7YP6hVgJZ8N6XgU58

QR Code:

Video Content Link:

https://www.youtube.com/playlist?list=PLzlh_31VWEFqOy0ZfEhyPBOVEZVcvNU0r

QR Code:

ADDITIONAL RESOURCES

Link: https://www.youtube.com/playlist?list=PLzlh 31VWEFrDhULahX3YAHzOKDaRZHLp

QR Code:

YouTube Videos Playlists for all Chapters

Chapter 1 Link: Https://Www.Youtube.Com/Playlist?List=Plzlh 31vwefr8ug6qr9xfgybqqdxurtbi

QR Code:

Chapter 2 Link: Https://Www.Youtube.Com/Playlist?List=Plzlh 31vwefrlhrrc3nzje3raivnlhaeo

QR Code:

Chapter 3 Link: Https://Www.Youtube.Com/Playlist?List=Plzlh 31vwefrqfouezv8pkugghxpsrlno

QR Code :

Chapter 4 Link: Https://Www.Youtube.Com/Playlist?List=Plzlh 31vwefqitzjm05ydcjaxl8z8mrtj

QR Code :

Chapter 5 Link: Https://Www.Youtube.Com/Playlist?List=Plzlh 31vwefrpzqqqsnvjfvqtf77rxojj

QR Code :

Chapter 6 Link: Https://Www.Youtube.Com/Playlist?List=Plzlh 31vwefobwb-Navnubgmhyncyavhw

QR Code :

Chapter 7 Link: Https://Www.Youtube.Com/Playlist?List=Plzlh 31vwefpeqtlsec9glteirst01xv-

QR Code :

Chapter 8 Link: Https://Www.Youtube.Com/Playlist?List=Plzlh 31vwefp20unkz260nt6nlyquybtn

QR Code :

Chapter 9 Link: Https://Www.Youtube.Com/Playlist?List=Plzlh 31vwefqoy0zfehypbovezvcvnu0r

QR Code :